OXFORD MONOGRAPHS ON GEOLOGY AND GEOPHYSICS

OXFORD MONOGRAPHS ON GEOLOGY AND GEOPHYSICS

1. DeVerle P. Harris: *Mineral resources appraisal: mineral endowment, resources, and potential supply: concepts, methods, and cases*
2. J. J. Veevers (ed.): *Phanerozoic earth history of Australia*
3. Yang Zunyi, Cheng Yuqi, and Wang Hongzhen (eds.): *The geology of China*
4. Lin-gun Liu and William A. Bassett: *Elements, oxides, and silicates: high-pressure phases with implications for the Earth's interior*
5. Antoni Hoffman and Matthew H. Nitecki (eds.): *Problematic fossil taxa*
6. S. Mahmood Naqvi and John J. W. Rogers: *Precambrian geology of India*
7. Chih-Pei Chang and T. N. Krishnamurti (eds.): *Monsoon meteorology*
8. Zvi Ben-Avraham (ed.): *The evolution of the Pacific Ocean margins*
9. Ian McDougall and T. Mark Harrison: *Geochronology and thermochronology by the $^{40}Ar/^{39}Ar$ method*
10. Walter C. Sweet: *The conodonta: morphology, taxonomy, paeoecology, and evolutionary history of a long-extinct animal phylum*
11. H. J. Melosh: *Impact cratering: a geologic process*
12. J. W. Cowie and M. D. Brasier (eds.): *The Precambrian-Cambrian Boundary*
13. C. S. Hutchinson: *Geological evolution of south-east Asia*
14. Anthony J. Naldrett: *Magmatic sulfide deposits*
15. D. R. Prothero and R. M. Schoch (eds.): *The evolution of perissodactyls*
16. M. Menzies (ed.): *Continental mantle*
17. R. J. Tingey (ed.): *Geology of the Antarctic*
18. Thomas J. Crowley and Gerald R. North: *Paleoclimatology*
19. Gregory J. Retallack: *Miocene paleosols and ape habitats in Pakistan and Kenya*
20. Kuo-Nan Liou: *Radiation and cloud processes in the atmosphere: theory, observation, and modeling*
21. Brian Bayly: *Chemical change in deforming material*
22. A. K. Gibbs and C. N. Barron: *The geology of the Guiana Shield*
23. Peter J. Ortoleva: *Geochemical self organization*
24. Robert G. Coleman: *Geologic evolution of the Red Sea*
25. Richard W. Spinrad, Kendall L. Carter, and Mary Jane Perry: *Ocean optics*
26. Clinton M. Case: *Physical principles of flow in unsaturated porous media*
27. Eric B. Kraus and Joost A. Businger: *Atmosphere-ocean interaction, 2nd edition*
28. M. Solomon and D. I. Groves: *The geology and origins of Australia's mineral deposits*
29. R. L. Stanton: *Ore elements in arc lavas*
30. P. Wignall: *Black shales*
31. Orson L. Anderson: *Equations of state for solids in geophysics and ceramic science*
32. J. Alan Holman: *Pleistocene amphibians and reptiles in North America*

TIME IN MILLIONS OF YEARS BEFORE PRESENT	ERAS	PERIODS	EPOCHS
.1	CENOZOIC	QUATERNARY	HOLOCENE
1.9			PLEISTOCENE
5		TERTIARY	PLIOCENE
25			MIOCENE
35			OLIGOCENE
55			EOCENE
65			PALEOCENE
140	MESOZOIC	CRETACEOUS	
210		JURASSIC	
250		TRIASSIC	
290	PALEOZOIC	PERMIAN	
320		PENNSYLVANIAN	
360		MISSISSIPPIAN	
410		DEVONIAN	
440		SILURIAN	
500		ORDOVICIAN	
550		CAMBRIAN	
4550	PRECAMBRIAN		

Pleistocene Amphibians and Reptiles in North America

J. ALAN HOLMAN

OXFORD UNIVERSITY PRESS New York
CLARENDON PRESS Oxford
1995

Oxford University Press

Oxford New York
Athens Auckland Bangkok Bombay
Calcutta Cape Town Dar es Salaam Delhi
Florence Hong Kong Istanbul Karachi
Kuala Lumpur Madras Madrid Melbourne
Mexico City Nairobi Paris Singapore
Taipei Tokyo Toronto

and associated companies in
Berlin Ibadan

Published by Oxford University Press, Inc.,
200 Madison Avenue, New York, New York 10016

Library of Congress Cataloging-in-Publication Data
Holman, J. Alan, 1931–
Pleistocene amphibians and reptiles in North America/
J. Alan Holman.
p. cm. — (Oxford monographs on geology and geophysics; no. 32)
Includes bibliographical references and index.
ISBN 0-19-508610-4
1. Amphibians, Fossil—North America.
2. Reptiles, Fossil—North America.
3. Paleontology—Pleistocene.
4. Animals, Fossil—North America.
I. Title. II. Series.
QE867.H65 1995 567'.6—dc20 94-32144

9 8 7 6 5 4 3 2 1
Printed in the United States of America
on acid-free paper

Dedicated to the Pioneers of Pleistocene Paleoherpetology

Walter Auffenberg
Bayard Brattstrom
Joseph A. Tihen

PREFACE

The Pleistocene epoch is characterized by climatic fluctuations and sea level changes wrought by advancing and retreating continental ice sheets over an interval that started almost two million years ago and ended only about ten thousand years before the present. This age saw the rise and spread of humans and a great extinction of the mammalian megafauna by the end of the epoch.

Recent major works have dealt with Pleistocene mammals in North America and the "why" of the mammalian extinction. Yet no comparable work deals with North American Ice Age amphibians and reptiles. This book attempts to rectify this situation by focusing on the North American Pleistocene herpetofauna and its unique relationships to the compelling events of the epoch.

The book begins with a general discussion of the Pleistocene and its effects on the physical and biological world. A discussion of dating methods and the nomenclature of established Pleistocene chronologies follows. Next a synopsis of herpetological families and genera recorded from the North American Pleistocene is presented. The section that follows deals with the fossil skeletal elements that are used in the identification of Pleistocene amphibians and reptiles and why these elements are utilized.

Next, the major North American herpetofaunas are detailed by geographic region. Information includes the location, age, and herpetological composition of each fauna. The last chapters of the book include a discussion of herpetological population adjustments and extinction patterns in the Pleistocene of North America and a consideration of the evolutionary "stasis" of the herpetofauna compared to the avian and mammalian faunas.

This book has been written at the "college level" of comprehension. It is hoped that it will be an important reference for anyone interested in Pleistocene flora and fauna, as well as Pleistocene events in general. It is also hoped that it will be of interest to modern herpetologists, ecologists, and evolutionary biologists. Nevertheless, I have tried to write the book with as little scientific jargon and with as few convoluted arguments as possible.

East Lansing, Mich. J. A. H.

CONTENTS

Pleistocene Amphibians and Reptiles in North America

1

INTRODUCTION

The Pleistocene epoch, or "Ice Age," is thought to have begun about 1.9 million years ago (Repenning, 1987) and to have ended about 10,000 years ago (Meltzer and Mead, 1983). The term Pleistocene was proposed by Charles Lyell (1839) on the basis that it contained more than 70 percent of Recent species of marine mollusks, principally from beds in southern Italy. This contrasted with the time period now known as the Pliocene, which was characterized by 40 to 70 percent Recent mollusk species and began about 3.1 million years before the Pleistocene. Later, Forbes (1846) correlated the Pleistocene with European glaciation, and the idea of a worldwide Pleistocene followed shortly after.

In 1948 the base of the Pleistocene was defined, in an edict by the eighteenth International Geological Congress, as the Italian Calabrian stage (marine) and its equivalent Villafranchian stage (terrestrial). This base recognizes as the starting point of Pleistocene events climatic deterioration in Italy rather than the appearance of the first continental glaciation. Since 1948 an attempt has been made to correlate the Italian Calabrian–Villafranchian sequence with other stratigraphically equivalent units worldwide.

The Pleistocene is characterized by glacial stages in which gigantic glaciers (ice sheets), moving down from the north, blanketed the landscape and interglacial stages in which the ice retreated. The glaciers were massive forces on the land, sculpturing the topography, carrying much sedimentary material, including huge boulders, over great distances. One can get an idea of the power of the ice sheets from the megablocks of the Canadian prairies, gargantuan masses of Cretaceous bedrock consisting mainly of bentonitic shales and sandstones, that weigh tens, even hundreds, of millions of tons. Megablocks up to 4 km long were transported by the ice over distances as much as 250 km (MacStalker, 1977).

The advance and retreat of ice sheets had a marked effect on the distribution of plant and animal life. Huge tracts of habitat were alternately obliterated and reopened several times. But other effects perhaps are not so well known. For instance, during glacial times, so much atmospheric water was bound up in the formation of the ice that sea levels worldwide lowered appreciably. On the other hand, during interglacial times, when water was melting from the ice sheets, sea levels rose, becoming higher than at present.

For biologists, the most interesting and controversial event that took place in the Pleistocene was the sudden, virtually worldwide, wave of extinction of large mammals and, to a lesser extent, birds that took place about 10,000 years ago. At least 200 mammalian genera became extinct, among them large herbivores, the carnivores that ate them, and the scavengers that fed on the remains of both.

Some of the striking kinds of endothermic animals that became extinct in North America were the giant condor, the dire wolf, the cave bear, the American lion (bigger than the modern African lion), the sabertooth cat, the giant ground sloth, the glyptodont, the giant armadillo, the capybara, the giant beaver (a form as large as a black bear), several taxa of horses, camels, tapirs, peccaries, mammoths, the American mastodont, and the giant bison (see Anderson, 1984).

Considerable discussion and debate about the extinction of the "mammalian megafauna" have taken place over the past decades (e.g., Martin and Wright, 1967; Martin and Klein, 1984; Stuart, 1991), with the main issue being whether the rise of human big-game hunters or changing environmental conditions were the major causes. But surprisingly few works have been written about the other classes of vertebrates and their part in the complex events of the Ice Age, and no summary works have been written about herpetofauna. This book, then, deals with the Pleistocene amphibians and reptiles of the North American continent (excluding Mexico) with emphasis on their ecological relationships and patterns of evolution and extinction. As it happens, the North American herpetofauna was quite stable during the Pleistocene, compared to the avian and mammalian fauna, which suffered familial, generic, and much specific extinction (Holman, 1991b). A compelling question, then, is "What attributes have allowed the herpetofauna to survive the stresses that caused dramatic extinctions in other vertebrate classes in the Pleistocene?" This question will be addressed after a detailed review of the North American Pleistocene herpetofauna.

We shall begin with a general discussion of the Pleistocene, including the effects of glaciation and deglaciation and some of the problems inherent in the interpretation of Pleistocene events. A discussion of dating methods and the nomenclature of established Pleistocene chronologies will be included.

2

THE PLEISTOCENE IN NORTH AMERICA

The Pleistocene epoch is believed to have begun about 1.9 million years ago. The North American Pleistocene is divided into stratigraphic units based on (1) glacial and interglacial ages and (2) North American Land Mammal Ages. This will be discussed in some detail later in this chapter. The period of time from the end of the Pleistocene through the present is designated the Holocene.

The most dynamic feature of the Pleistocene in North America was the presence of glaciers and their advances and retreats. In the classic interpretation there were four advances and four retreats, the last retreat occurring near the end of the epoch. These movements occurred with the periodic growth and degeneration of two giant ice masses. The largest of these was the Laurentide Ice Sheet, which extended from Nova Scotia and the northeastern United States across the continent to western Canada. A smaller mass, the Cordilleran Ice Sheet, blanketed the mountain ranges of the Northwest from Montana and Washington up to the Aleutian Islands (Matsch, 1976).

The deepest penetration of this glacial field took place during the Illinoian glacial stage, when the Laurentide Ice Sheet penetrated as far as southern Illinois, Indiana, and Ohio. During the Wisconsinan (last glacial age) the maximum penetration of the ice sheet was to central Illinois, Indiana, and Ohio at almost identical latitudes (Fig. 1), following the weather pattern of the time.

Effects of Glaciation and Deglaciation

Several aspects of the glacial cycles undoubtedly had a marked influence on biotic communities and their amphibian and reptile populations. Some of these aspects are summarized in this section.

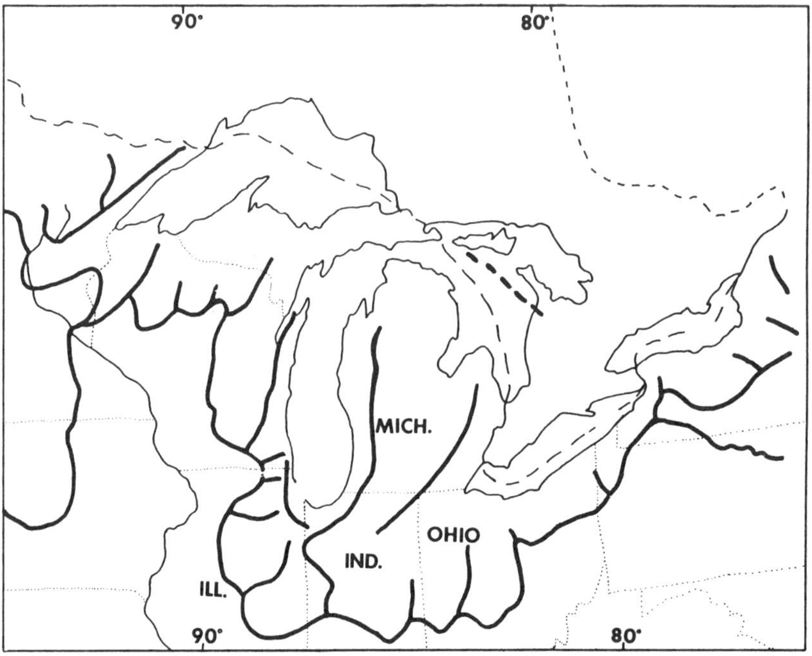

FIGURE 1. Maximum extent of the Laurentide ice sheet in the late Wisconsinan in the Great Lakes region showing the principle lobes and sublobes 21,000 to 20,000 years ago. [From Holman, 1992a]

Community Obliteration

When the glaciers advanced, vast areas of habitat were obliterated. For example, in the late Wisconsinan, about 20,000 ybp, the ice sheet smothered huge areas of herpetological habitat as well as restricting this habitat in the southern unglaciated parts of Illinois, Indiana, and Ohio (Fig. 1).

Succeeding advances occurred at intervals of about 1,000 years, with a final advance extending to northern Indiana and Ohio about 15,000 ybp. After about 14,800 ybp (Fig. 2), the ice began its final retreat in this area, reexposing habitat for the reinvasion of plants and animals. The movement of amphibian and reptile species into these newly exposed areas established the modern herpetofauna in Michigan and in northern Illinois, Indiana, and Ohio.

Topographic Changes

The glacial advances and retreats radically altered the topography of the areas that they passed over. The thickness of the ice sheets varied, but it has been estimated that the average thickness was about 2 km and that the maximum thickness of some may have been 3 km or more (Flint, 1971). Valleys and

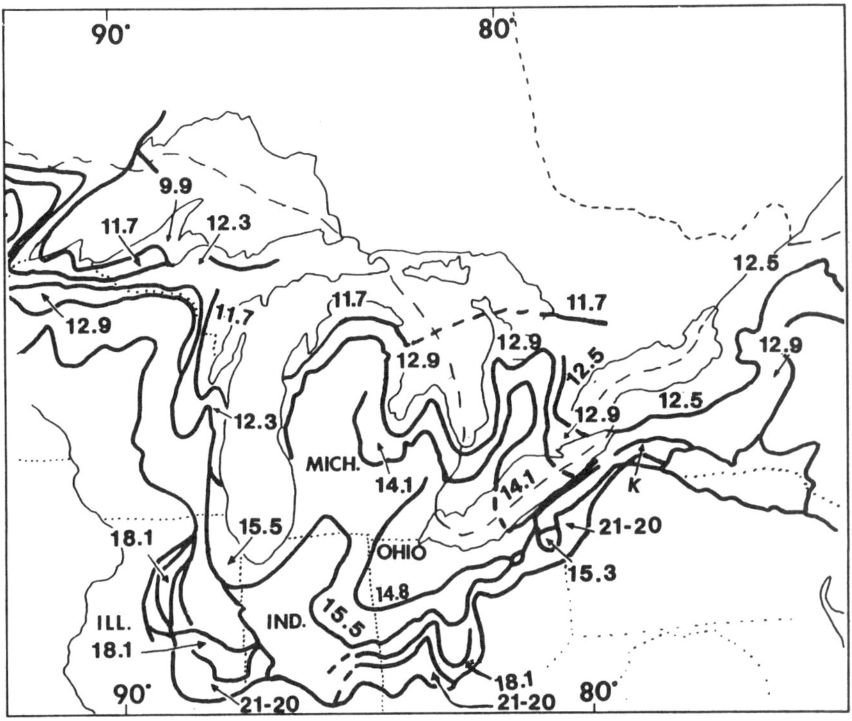

FIGURE 2. Ice margin positions in the Great Lakes region in thousands of years ago. [From Holman, 1992a]

ridges, various types of small hills, lakes, streams, swamps, and bogs were all produced by glacial activity. Some of these features were large—the Great Lakes among them—and others were relatively small. All provided a variety of habitats for plants and animals.

Climatic Changes

Of the many changes that accompanied glacial cycles, climatic change has received considerable attention. The classic idea of alternating cold glacial and warm glacial and interglacial climates (e.g., Dunbar, 1949) is now considered to be oversimplified. Modern evidence indicates that climates were cold in proglacial areas, such as southern Michigan and Wisconsin, but in the central and southern United States the climate is believed to have been more equable than today, with warmer winters and cooler summers (Lundelius et al., 1983). This theory, termed the "Pleistocene climatic equability model," will be discussed in detail in Chapter 7. It is an extremely important consideration relative to mixtures of northern and southern herpetological species in Pleistocene faunas.

Vegetational Changes

Climatic change wrought by the glaciers greatly affected biotic communities south of the glaciated areas. The classic idea is that major vegetational associations were pushed southward en masse by the advancing ice. For example, it was believed that in the eastern United States during glacial periods a barren tundra association existed in a broad band immediately south of the glacier, followed by a broad band of taiga grading into temperate deciduous forest that penetrated deep into the Southeast (Holman, 1975).

The modern synthesis, however, is that during much of the Pleistocene a cold climate and tundra or taiga vegetation existed in the proglacial areas near the ice mass, but for the most part the vegetational communities in the central and southern United States existed as a mosaic of plants that remained "in place" and certain northern invading species (Graham and Lundelius, 1984).

This idea has grown from the concept that plant (and animal) species reacted individually rather than as collective units to glacial changes, and the "mosaic communities" just referred to above would be able to exist in the equable climates south of the proglacial areas (Graham and Lundelius, 1984). The classic concept is often referred to as the "stripe hypothesis," the modern as the "plaid hypothesis."

Sea and Lake Level Changes

During the Pleistocene so much global water was bound up in the ice sheets that sea levels fell. In Florida, the peninsula enlarged considerably as the sea withdrew during glacial times (Cooke, 1945) and was so encroached by the rising sea during the interglacials that it was cut off from the mainland, with only its highest central portion (the "Ocala Island") left standing above the sea. In the Great Lakes region the lakes themselves rose and fell, and such rivers as the Illinois and the St. Lawrence may have become inland seaways (Dorr and Eschman, 1970).

Lag Time in the Development of New Communities

Melting ice from retreating glaciers leaves a mass of sterile mud, silt, sand, and gravel in its wake. This material must be invaded by pioneer organisms in order for biotic communities to reestablish themselves. Glacial geologists, camping upon such sterile sediments ahead of modern glaciers, have reported that *Dryas,* willows, spiders, and voles are the early pioneering species in such glacial sediments (Grahame J. Larson, pers. comm. 1992).

Certainly there must have been a lag time between the retreat of the Pleistocene ice sheets and the reestablishment of fully developed, stable plant communities. If small glacial oscillations occurred between the larger glaciations, it would seem that the lag time could be hundreds of years.

A gap of several scores of miles between the northernmost records of late

Wisconsinan vertebrates in Michigan and the glacial fronts to the north indicate a "tension zone" (Holman, 1991a), where the lag time before communities could accommodate large vertebrates must have been considerable. Odd assemblages of smaller animals and plants might have existed in disequilibrium in such areas.

Some Major Questions

Many scientists think that the Pleistocene is the most intriguing unit of geological time. Although it ended only about 200 human generations ago, many compelling questions exist.

Cause

An explanation for the cause of glacial ages has been pursued ever since Louis Agassiz (1807–1873) proposed the theory of the Great Ice Age. However, a cohesive theory—such as those explaining evolution (natural selection) and continental drift (plate tectonics)—has not yet emerged to deal with the cause of Pleistocene (or earlier) glacial events. Such a theory may lie in hypotheses related to cyclic changes in the atmosphere. These changes have been attributed variously to fluctuations of solar output, irregularities in Earth's orbital pattern and rotation, surging activities of the Antarctic ice sheet, volcanic periodicity producing sunlight-blocking dust, and variations in Earth's magnetic field (Matsch, 1976).

It also has been suggested that major and minor ice cycles are each caused by different factors. For instance, major cycles might be caused by changes in Earth's orbit, and minor cycles, or "Little Ice Ages," might be caused by solar variations (Imbrie, 1977). Atmospheric studies generated by emerging technologies perhaps will generate a unified working theory for the cause of the ice ages.

Reality of the Holocene

The Holocene represents the 10,000 or so years elapsed since the end of the Pleistocene epoch. But is the Pleistocene actually over? In Britain it is common to use the term "Flandrian" for the Holocene, with the implication that the Flandrian represents yet another temperate stage of the epoch (Stuart, 1982). Since there is no scientific evidence that the causes for the climatic rhythms of the last 1.9 million years have subsided, it would seem reasonable to assume that another glacial age lies in the future.

Evidence is that the last interglacial age, the Sangamonian (Ipswichian in Britain), lasted 10,000 years, and that the last glacial age, the Wisconsinan (Devensian in Britain), lasted 100,000 years (Stuart, 1982). It has been 10,000 years since the end of the Wisconsinan/Devensian, and if the cycle is repeating itself, we should expect another ice age very soon!

Extinction of the Mammalian Megafauna

In contrast to earlier mass extinctions, such as those at the end of the Permian and the end of the Cretaceous, the extinction at the end of the Pleistocene mainly affected one group, the mammalian megafauna, or large land mammals, ranging in size from piglike peccaries to elephantlike mastodonts and mammoths (Martin, 1984). One of the most intensively studied of the Ice Age questions is the "why" of this extinction.

Considered at the generic level, the extinction of mammals of peccary size (about 44 kg adult weight) or larger was very striking, but these extinctions varied in different regions. In Australia 94 percent (15 of 16 genera) were lost, and in South America about 80 percent (46 of 58 genera) became extinct. North America lost 73 percent (33 out of 45) of its large mammalian genera. But Europe suffered only a moderate extinction of 29 percent (7 of 24 genera), and in Africa, south of the Sahara, only about 5 percent (2 of 44 genera) died out (Klein, 1984; Martin, 1984).

The modern consensus is that the large, slow-breeding herbivore species that were few in number and needed the most living space were the most vulnerable. It is also thought that the extinction of the herbivores led to the extinction of the carnivores that preyed upon them and the scavengers that cleaned up after the carnivores.

Owen-Smith (1987) has stressed the importance of the largest herbivores, called megaherbivores, as they relate to environmental stability in modern Africa, pointing out that the feeding activities of elephants and rhinos quickly change wooded savanna to open short-grass areas dominated by rapidly regenerating plants that provide food for smaller herbivores. Elimination of elephants and rhinos leads to massive environmental changes, and the extinction of megaherbivores at the end of the Pleistocene may have caused changes from grasslands to forests.

Very few genera of small mammals became extinct in the late Pleistocene, but a rather large number of species were lost (Kurten and Anderson, 1980). Nevertheless, it has been pointed out that more extinct small mammal species may have been named than actually existed (Stuart, 1991). This might also be said about some large mammalian species such as extinct *Equus* (horse) species from North America.

Birds suffered much less familial and generic extinction at the end of the Pleistocene. However, several extinct genera of birds are known from the Pleistocene of Australia (Murray, 1984) and North America (Steadman and Martin, 1984). Late Pleistocene extinction of reptile, amphibian, and fish taxa was less severe than in mammals and birds. Extinction in North American herpetological taxa will be discussed in detail in Chapter 8.

The reasons for Pleistocene extinctions has been a popular topic in scientific circles since 1812, and by the mid-nineteenth century most possible explanations for these events had been discussed (Grayson, 1984). At present, there are two major hypotheses: (1) the large mammals were unable to adapt to climatic changes at the end of the Pleistocene or (2) prehistoric peoples hunted them to extinction.

Scientists who favor the climatic hypotheses have pointed out several events that could have led to the extinction of the mammalian megafauna. It has been shown that the vast, unglaciated, grass- and herb-rich area in northern Eurasia and Alaska (the "mammoth steppe") gave way to forest at the end of the Pleistocene (Stuart, 1991), thus eliminating an entire biome and many of its large mammals—including the resident megaherbivore of the biome, the woolly mammoth.

Graham and Lundelius (1984) have pointed out that Pleistocene terrestrial vertebrate communities had co-evolved to such an extreme that each of the large herbivorous mammalian species modified the vegetation in a way that benefited the others. They suggest that when climatic and vegetational changes occurred at the end of the Pleistocene, each species responded individually, thus destroying the previous balance, and causing much extinction.

Slaughter (1967) and Axelrod (1967) have suggested an "out-of-step mating hypothesis." This would have resulted when areas outside the tropics experienced changes from equable to harsh climates at the end of the Pleistocene. They reason that the reproductive cycle of large mammalian species with long gestation periods would have been in step with existing Pleistocene climatic patterns. For example, breeding in the fall would have been timed so that the young were born in the spring when new green vegetation was available. But the sudden onset of harsh Holocene climates would mean that the young would be born without this new supply of food. It was unlikely, they believe, that genetic patterns for in-step breeding would have evolved fast enough to save the large ungulates and their dependent carnivores.

The most rigorous objections to the various climatic hypotheses are voiced in two questions: (1) Why were the large mammals that became extinct unable to find refugia with resources that met their needs? (2) Why were there no major extinctions in previous Pleistocene transitions from warm to cold or, conversely, from cold to warm periods such as occurred at the end of the Wisconsinan?

Proponents of the hypothesis that paleohumans hunted the large mammals to extinction (e.g., Martin, P., 1973, 1984) do not think that the Pleistocene–Holocene transition was any different from other glacials and interglacials. They believe that humans arriving in North America and then migrating into Central and South America brought advanced hunting technologies that devastated the large mammals in the newly invaded continents.

They reason that the large mammals had never been confronted by human hunters and, thus unwary of humans, were easy to kill. They also believe that the humans were unused to such easy kills and that they greatly overexploited their prey. In response to the criticism that remains of extinct species are rarely found in association with human artifacts, Martin (1973) advanced a "blitzkrieg hypothesis" purporting that human predation on the fauna in the New World came so swiftly and completely that little evidence was left behind.

Chronology of the North American Pleistocene

Geological events are put into a temporal sequence using two techniques, relative and absolute dating. Relative dates relate one event to another in a given sequence of geological events, absolute dates indicate in years the time that has passed since a specific geological event occurred. Following are some of the common chronological methodologies used in studies of the North American Pleistocene amphibians and reptiles.

Relative Dating

A relative chronology in ancient rocks is established mainly through the geological principle of **superposition**, whereby younger beds lie on top of older ones, and on the principle of **faunal succession**, that fossil assemblages change from older to younger rocks. Some of the common situations where North American Pleistocene amphibians and reptiles can be found in a relative chronological context are considered next.

Deposits Associated with River Terraces.

When rivers change their courses laterally and deepen their channels, the older deposits form a series of terraces above the the present river. The sedimentary layers in each of these terraces are arranged in a chronological series from bottom to top. As the river continues to move laterally, progressively younger series of terraces may form. Thus the oldest deposits are found farthest from the river, in the higher terraces, and the youngest deposits are found in the river bed. Many factors of course can complicate this picture, and river terrace stratigraphy may be quite complex. Amphibian and reptile fossils are found in river terraces in Florida, Texas, and in the Great Plains states and may be dated, in part, on the basis of their stratigraphic occurrences in these terraces.

Deposits Associated with Marine Terraces

Sometimes Pleistocene vertebrate assemblages have been assigned to specific interglacial or glacial ages based on their association with marine terraces. However, Pleistocene marine terrace deposits, especially early ones, can be difficult to interpret when tectonic events have caused displacement of the land. Thus caution should be used in correlating Pleistocene marine terraces with one another on the basis of their altitudinal relationships alone.

In some instances, nonmarine features have been incorrectly interpreted as marine terraces. In northern Florida, an eroded surface of the Ocala Limestone was mistaken for a Sangamonian marine terrace, and herpetological fossils from several localities in solution features directly below the "terrace" were incorrectly considered to be of Illinoian age (e.g., Holman, 1959a).

Deposits Associated with Glacial Gravels or Tills

Sedimentary deposits above and below glacial gravels or tills are usually attributed to interglacial ages, especially if the fauna indicates a warm or temperate interval. For example, if a deposit in central Illinois is found above definitive Illinoian till and below definitive Wisconsinan till (superposition) and contains a temperate fauna (faunal succession), one would assume that the herpetofauna within would represent the Sangamonian interglacial age. Herpetofaunas have been identified in such situations in several areas in Illinois and in the Great Plains.

Deposits Associated with Glacial Lake Deposited Sediments

In proglacial areas of the Great Lakes Basin, glacial lake sediments mark changes in the levels of the ancient Great Lakes. Vertebrate bones, including those of toads, have been found below Great Lakes glacial lake sediments in water-well diggings (Holman, 1988), but other herpetological finds in these situations have been rare.

Glacially Derived Kettle Bogs and Shallow Basins

Kettle bogs and shallow basins are the most common repositories of fossil amphibians and reptiles in northern proglacial areas. When the Wisconsinan glaciers retreated they left behind many depressions (kettles) that filled with glacial meltwater. Many of the smaller kettles and basins eventually filled in and many large ones are still in the process. The stratigraphy of these features usually results from a successional process and is usually depicted by five zones (from bottom to top).

Zone 1 is composed of unfossiliferous glacially derived clastic sediments, such as sands and gravels. **Zone 2** is often a sterile, grayish blue clay that was produced chemically and physically from the parent glacial material. **Zone 3** is usually a grayish-brown, colloidal, shelly marl containing many fossils, mainly the detrital and skeletal remains of plants and animals that accumulated in the well-aerated biotic stage of the ecological succession of the feature. A square meter of this material may contain hundreds of freshwater shells and occasionally frog, turtle, and snake remains. **Zone 4** normally consists of dark organic peat or muck that represents a later seral stage of the feature where mats of aquatic vegetation formed over the surface. Occasionally, frog, turtle, and snake remains occur here. **Zone 5** consists of recently derived humus and topsoil.

The herpetological fossils found in such deposits are usually not numerous because the biotas of such deposits are usually those indigenous to cool or cold proglacial areas. Carbon 14 dating on wood, other plant remains, or bones may tie down the age of amphibians and reptiles found in situ. Care needs to be taken in assigning dates to amphibian and reptile bones based on the absolute dating of mastodont or mammoth bones collected in kettles or basins. The probability is that the mastodonts or mammoths fell through a covering mat of vegetation and drowned or became stuck and were killed by humans when the feature was in the "quaking bog" stage, long after the herpetological species were preserved as fossils.

Pleistocene Solution Features

Most large North American Pleistocene herpetofaunas are found in caves, sinkholes, and other solution features. The most common type of cave or sinkhole forms as a result of the solution of limestone by somewhat acid groundwater. These features usually start as small cavities along weak areas in limestone strata below the water table; over many thousands of years they may reach a very large size. Mammoth Cave in Kentucky has over 200 km of interconnecting passages (Suttcliffe, 1985). When the water table lowers, as occurs in time of drought, caves may collapse as the roof falls in. In Florida, sinkholes commonly form by this process.

Unfortunately, establishing a relative chronology in solution features can be difficult. The biggest problem is that Pleistocene infillings are often washed out later by stream action that deposits new sediments on the cave floor and leaves remnants of the Pleistocene deposits on the roof, walls, and especially on ledges within the cave. Thus, the older deposits in a cave setting are often higher than the younger ones, an **inverted chronological sequence**.

Typically, the sediments on the floor of a Pleistocene cave contain the skeletal remains of modern bats and other fauna that live in the cave. Chimney caves, whose only openings are vertical shafts, often act as natural traps, and skeletons of dogs, farm animals, and sometimes humans recently fallen into the cave are mixed into the bony potpourri. Interpreting the cave floor deposits might become quite confusing if extinct Pleistocene vertebrates are found mixed in with the bones of modern bats or dogs, as well as beer cans. This could occur when episodes of extreme cave flooding wash fossils out of wall or ledge deposits, but fortunately this is rare.

Paleontologists thus look for vertebrate fossil-containing sediments on the cave walls and ledges, to find extinct Pleistocene animals in situ. They utilize the principle of faunal succession as well as superposition (or reversed superposition) in establishing a relative chronology.

Sometimes deposits in caves form as identifiable stratigraphic layers. Such a process, fortunate for the paleontologist, takes place if the cave roof opens and closes several times during its history. If the layers form on a distinct ledge, the sediments washing in from outside build up normally, that is, with the oldest sediments on the bottom of the ledge and the younger layers toward the top.

Outside the cave, weathering and erosion of sediments and temporal changes in vegetation, soil, and fauna can produce sedimentary layers of different color or texture through time. Thus the layers of sediments on the ledge may have different color, texture, and composition. A massive forest fire might eliminate the species of a mature forest community and replace them with pioneer species of plants and animals. Or over an extended period of time, the climate itself might change. Thus sediments on a cave ledge may differ depending on when the openings in the roof occurred. The contacts between these layers then indicate periods when the roof was plugged and sedimentation on the ledge ceased.

Some of the early assumptions about the occurrence of Pleistocene amphibians and reptiles in caves and sinkholes have changed. Because many cave sediments contain herpetological species that would not be ecologically compatible with the present-day mammalian species (see Lundelius et al., 1983), workers have tended to disregard the herpetological specimens in the belief that they were modern intrusions in the sediments. This reasoning has been based on paleoclimatic interpretations of the mammalian component: the presence of a few boreal species of mammals led to the conclusion that the climate must have been much too cold for present-day herpetological species to have been associated with the mammals.

Actually, it is not difficult to recognize intrusions of modern herpetological species into these sediments. True intrusive forms are almost always burrowers such as mole salamanders of the genus *Ambystoma* or anurans such as *Scaphiopus*. The intrusive forms usually consist of suspiciously complete skeletons, and they are usually of a different color and density than those of the true fossils of the deposit.

Another assumption formerly made, especially in interpretation of southeastern North American caves (e.g., Holman, 1967), was that the presence of ecologically incompatible forms indicated a heterochronic deposit, that is, it represented a mixture of bones from different time periods. This reasoning undoubtedly grew out of misinterpretations of earlier writers (e.g., Sellards, 1916; Cooke, 1945) that solution features in the Southeast contain sedimentary layers with mixtures of Miocene, Pliocene, or Pleistocene fossils or mixtures of fossils of various Pleistocene glacial and interglacial ages.

The fact is that most solution deposits in the Southeast (with the exception of stream runs from springs) contain faunas that represent specific glacial or interglacial ages (Webb, 1974; Holman, 1976). The only fossils that regularly occur mixed with fossils of other ages are the readily identifiable invertebrate fossils redeposited from the limestone walls of the caves.

The reason for this reflects the basic paleontological principle that for fossilization to occur there must be immediate burial. The sequence is (1) an animal falls into pit traps or is carried into solution features by raptors or carnivores, and (2) rapid sedimentation, caused by a local rainstorm or flooding leads to the burial and fossilization of the bones. Those paleontologists who have explored seemingly identical sinkholes and caves are well aware that some contain vertebrate fossils while others do not. It is logical to assume that the sedimentary layers in these features produced by gradual sedimentary process would be mainly unfossiliferous.

Faunal and Floral Associations

Establishing the age of certain North American vertebrates based on their association with other key forms is possible, but there are complicating factors. For instance, one might reasonably assign a very late Wisconsinan age to turtle remains associated with mastodonts or mammoths in Michigan, even in the absence of radiocarbon dates. This is because the many radiocarbon dates on other Michigan proboscideans span a short time period, from about 12,500 to 10,000 years ago, in the very latest Wisconsinan. On the other hand, proboscideans in Nebraska and Kansas span a wide range of Pleistocene time. Thus, in poorly stratified or undated proboscidean sites, it would be difficult to assign associated turtle bones to any specific Pleistocene age.

One of the paradoxes in North American Pleistocene studies is that pollen diagrams often conflict with information derived from the study of associated vertebrate faunas, and in particular when compared to associated Pleistocene herpetofauna (Holman and Richards, 1993). The common interpretation is that pollen indicates cooler conditions than indicated by the herpetofauna. This could be attributed to the mosaic vegetation of the Pleistocene and the fact that some northern trees, such as pines, produce much more pollen than others.

Absolute Dating

Absolute dating assigns a specific time frame to the chronological events that are based on studies of the superposition of sediments and the faunas within them. At present there are several methods of absolute dating of Pleistocene events. Dates based on a time scale derived by counting cyclic events such as the formation of varves or tree rings have been successfully used in some northern regions (varves) or areas where trees live to a great age, such as Florida (cypress), Arizona (bristlecone pine), or Washington (giant redwoods), but these techniques have not been important in studies of North American Pleistocene vertebrates.

The carbon-14 method (radiocarbon dating), the most commonly used absolute method for dating North American Pleistocene vertebrate sites, is accurate to about 50,000 years ago. Radiocarbon dates are based on the fact that plants incorporate small amounts of the unstable radioactive isotope carbon-14 in their tissues. Since plants are the base of the food chain, carbon-14 is incorporated in all the higher trophic (feeding) levels as well. When any of these organisms die, carbon no longer enters the dead tissues and the residual carbon-14 is lost through radioactive decay.

If the amount of carbon-14 in a sample of previously living tissue is measured, the time since the death of the organism may be calculated. Because radioactive decay decreases the carbon-14 by one-half at regular intervals, there is usually not enough left to measure after about 50,000 years. Many kinds of organic material may be used to obtain radiocarbon dates, but the most common tissues used in North American Pleistocene vertebrate assemblages are bone and wood. Historically, bone has been used less than wood, but in the last twenty years, because of better techniques, bone dates have become common.

Two components of bone have been used to obtain radiocarbon dates: collagen and bone apatite. Collagen is by far the most dependable, although the tendency for it to weather out of the bone in certain situations may create problems. Bone apatite has been used if no collagen remains, but this method is of questionable accuracy because carbon in the groundwater may have replaced the original carbon in the apatite. The recent dating of bones generally is much more accurate than dating made a decade or two ago because methods of extracting collagen and removing organic contaminents have become more sophisticated. In fact, a former rule of thumb was that because of these contaminents, carbon-14 dates on bone were usually about 2,000 years younger than those on wood from the same site.

The use of fossil wood for dating can create pitfalls in certain situations. Pleistocene wood usually has such a high organic content that it may attract organisms such as rootlets and mold—especially if the fossil wood lies near the surface. Abnormally young dates would be obtained in instances where carbon has been added to the wood. In kettle bog sites, where fossil mastodonts and mammoths are often found, the probability is that the animals sank into older sediments containing wood in situ. Thus, bone dates from mastodonts and those from wood often conflict. On the other hand, wood dates in these situations are likely to be consistent with those of the amphibians or reptiles of the site.

Methods using the decay of potassium to argon (K-Ar) have been used to date Pleistocene sites in areas with potassium-rich volcanic lavas. These dates go back much further than 50,000 years. Several states in the Great Plains region have clearly definable layers of volcanic ash, sometimes several feet thick. These have been sources for potassium-argon dates in Pleistocene faunal studies and have led to new interpretations of the age of various Pleistocene faunas (see papers in Woodburne, 1987). In areas where there has been little or no volcanic activity, a technique based on the change of uranium to thorium has been used. Data obtained using this method can also go well beyond the 50,000-year limit of the carbon-14 method. However, the accuracy of the uranium-thorium method has been questioned because in certain instances the dates do not agree with the other stratigraphic and faunal evidence.

Pleistocene Glacial and Interglacial Ages

Early geologists in North America recognized that glacial deposits and landforms existed far south of existing glaciated areas, and they correctly surmised that these features indicated former cold climates. As these features were mapped and stratigraphic studies were made, it was found that some sections contained weathered zones of organic soils and plant remains between layers of glacially-derived sediments. These astute early workers suggested that these zones represented a nonglacial environment, and thus that the glaciers must have advanced and retreated several times.

Soon after the turn of the nineteenth century, four major drift sheets were recognized, each of which was distinguished from the others on the basis of evidence indicating an interglacial environment. The classic glacial and interglacial ages recognized in North America (oldest at the bottom) are:

Wisconsinan glacial age
Sangamonian interglacial age
Illinoian glacial age
Yarmouthian interglacial age
Kansan glacial age
Aftonian interglacial age
Nebraskan glacial age

The end borders of all of the glacial deposits are located in close proximity, and it has been suggested that each was formed under about equally severe climates. The Nebraskan and Kansan glacials are the most obscure because they have been exposed to weathering and erosion for many years and because the effect of later glaciations, especially the Wisconsinan, obliterated much of the record. The Wisconsinan glacial, the most recent, presents the most detailed record.

It has often been suggested that Pleistocene glacial events worldwide were much longer than the interglacials. It is believed that the Sangamonian interglacial (British Ipswichian) lasted about 10,000 years, but that the Wisconsinan glacial age (British Devensian) lasted about 100,000 years (Stuart, 1982). Age estimates of early glacial and interglacial ages are more obscure.

Vertebrate faunas are best known from late Wisconsinan deposits, moderately well known from earlier Wisconsinan, Illinoian, and Sangamonian deposits, and least well-known from earlier Pleistocene glacial and interglacial deposits. In the sections that follow, amphibians and reptiles will not only be related not only to glacial and interglacial events, but to Pleistocene units called "Land Mammal Ages."

Pleistocene Land Mammal Ages

Pleistocene Land Mammal Ages are based on changes in the North American mammalian fauna reflecting evolution, extinction, and dispersal (Lundelius et al., 1983). In this conceptual scheme, the first appearance of critical mammalian taxa is considered quite important. Large mammalian taxa with a wide range are often utilized for these interpretations, but of late, the evolution and dispersal of smaller taxa, such as rodents, have been considered important (Repenning, 1987). Significant progress has been made recently in comparing contemporaneous but geographically widely separated faunas based on potassium-argon and radiocarbon dating. The two North American Land Mammal Ages have been named the "Irvingtonian" and the "Rancholabrean." There is some overlap between the mammalian faunas of the late Pliocene Blancan Land Mammal Age and the early Pleistocene Irvingtonian that have caused a few workers to argue that the late Blancan should be considered part of the Pleistocene. But we will follow the majority of workers (in Woodburne, 1987) who consider all of the Blancan to be Pliocene.

Irvingtonian

The Irvingtonian Land Mammal Age was defined by Savage (1951) based on mammalian fauna from a gravel pit southeast of Irvington, Alameda County, California. It began about 1.9 million years ago and ended about 400,000 years ago, with the arrival of bison (*Bison*). Among the important mammals that first appear in the Irvingtonian are the mammoth (*Mammuthus*), modern horse (*Equus*), sabertooth cat (*Smilodon*), and hare (*Lepus*). It is assumed that many of these animals arrived by way of the Bering land bridge between Eurasia and North America. Several South American immigrants, such as the giant ground sloth (*Eremotherium*) and the huge armadillo (*Holmesina*), also reached North America during this time, but via the Central American land bridge.

Repenning (1987) has divided the late Hemphillian (middle Miocene) through the Holocene Land Mammal Ages into a series of ten stages involving microtine rodents. These stages reflect the invasion of advanced microtines from Asia into North America and the extinctions resulting from competition with more primitive local rodents. These events for the most part are dated by the potassium-argon method using ash beds that occur in stratigraphic relationship to the microtines. The microtine events are (after Repenning, 1987):

Irvingtonian I (microtine event 7 to event 8)	1,900,000 to about 900,000 ybp
Irvingtonian II (microtine event 8 to event 9)	900,000 to 400,000 $\pm$ 25,000 ybp

Rancholabrean

The Rancholabrean Land Mammal Age, also defined by Savage (1951), is named for the famous Rancho La Brea fauna in Los Angeles County, California. It began about 400,000 years ago and ended at the end of the Pleistocene about 10,000 years ago. The Rancholabrean is characterized by the Eurasion immigrant *Bison,* by the presence of many genera of large mammals that became extinct at the end of the Pleistocene, and by many extant species of small mammals, especially rodents and carnivores. Among other important Eurasian immigrants during this time were moose (*Alces*) and humans (*Homo*). Rancholabrean microtine rodent events are (after Repenning, 1987):

Rancholabrean I	
(microtine event 9 and event 10)	400,000 ± 25,000 to 150,000 ± 25,000 ybp
Rancholabrean II	
(microtine event 10 to the present)	150,000 ± 25,000 ybp to present

3

THE NORTH AMERICAN HERPETOFAUNA: LATE CRETACEOUS THROUGH PLIOCENE

The Pleistocene and Holocene North American herpetofauna is a product of past events. Although fossil records in general have many gaps, ancient herpetofaunas are much better known in North America than they are in many continents, and they give us a fairly good picture of changes that occur at the family and generic level. With the exception of a major radiation of advanced snake groups that occurred in the Miocene, the North American herpetofauna did not change nearly as much in the Cretaceous through Pliocene time span as did the avian and mammalian faunas.

In the following temporal accounts, many families and genera will be mentioned that also occur in the Pleistocene. Thus the reader is referred to the "Bestiary" (Chapter 4) that follows, where these taxa are systematically listed, discussed, and sometimes figured.

Late Cretaceous

An extensive fauna of salamanders, anurans, turtles, lizards, and crocodilians, as well as a few primitive snakes, coexisted with the dinosaurs in the North American late Cretaceous (Estes and Baez, 1985). The North American late Cretaceous herpetofauna is surprisingly modern at the family level, but there are no uncontested modern genera known. As a whole, the herpetofauna has its greatest affinities with North America and northern Eurasia (Holarctic region) and lesser affinities with Africa and South America. Several extinct endemic families exist.

Two modern North American eellike salamander families, the amphiumas, or "congo eels," of the family Amphiumidae and sirens of the family Sirenidae, are known from the late Cretaceous, as are two extinct endemic families, the Batrachosauroididae and the Scapherpetontidae. The modern North American anuran pelobatid and spadefoot family Pelobatidae occurs along with the modern Asian disk-tongued toad family Discoglossidae, as well an an extinct anuran family, the Palaeobatrachidae.

Turtles became quite diverse in the late Cretaceous of North America. Living North American families include the Central American river turtle family Dermatemydidae, the softshell turtle family Trionychidae, the snapping turtle family Chelydridae, and possibly the mud and musk turtle family Kinosternidae. These occur with the Asian family Macrobaenidae, some side-necked genera of the family Pelomedusidae found in marine sediments, and the extinct Baenidae (Estes and Baez, 1985).

The modern tropicolitan crocodile family Crocodylidae is present in the late Cretaceous of North America and is represented by several forms including a gaviallike animal. The modern North American family Alligatoridae is represented by five genera, one of which appears to be caimanlike (Estes and Baez, 1985).

Lizards of the late Cretaceous of North America are diverse and have a very modern aspect at the family level. Modern North American families are represented by the whiptail-racerunner family Teiidae, the anguid family Anguidae, the xenosaur family Xenosauridae, and the beaded-lizard family Helodermatidae. Two modern families with tropicolitan affinities are present, the skink family Scincidae and the varanid family Varanidae. One extinct lizard family, the Necrosauridae, is known. Snakes of the late Cretaceous in North America are represented by a single extinct genus of the modern pipesnake family Aniliidae and an indeterminate snake of the family Boidae.

In summary, the late Cretaceous herpetofauna of North America (excluding dinosaurs and other extinct orders) consists of eighteen living families and six extinct ones. Amphibians are represented by four living and three extinct families and reptiles are represented by fourteen living families and three extinct ones. A few living genera have been reported from the late Cretaceous, but these have all been seriously questioned.

Paleocene

Several families of salamanders, anurans, turtles, lizards, and crocodilians survived the dinosaur extinction episode in the latest Cretaceous and some new modern families were added. Moreover, uncontested modern genera of amphibians and reptiles show up in the Paleocene for the first time. The North American Paleocene herpetofauna continued to have its greatest affinities with the Holarctic region.

Three modern families of North American salamanders appear for the first time in the Paleocene. The hellbender and giant salamander family Cryptobranchidae is represented by the living genus *Cryptobranchus*, the olm-mudpuppy-waterdog family Proteidae is represented by the living genus *Necturus*, and the mole salamander family Ambystomatidae is represented by extinct genera. The amphiuma family Amphiumidae and the siren family Sirenidae are late Cretaceous survivors and are represented by extinct genera. The extinct salamander families Batrachosauroididae and Scapherpetontidae are also Cretaceous carryovers. Anurans are represented by the disk-tongue toad family Discoglossidae (also late Cretaceous carryovers), but the modern burrowing toad family Rhinophrynidae makes its first appearance in the Paleocene.

The turtle families of the Paleocene are the same as they were in the late Cretaceous, except Baenidae and the side-necked turtle family Pelomedusidae dropped out and the pond-river turtle family Emydidae made its first appearance. The modern softshell genus *Trionyx* (Trionychidae) also appeared for the first time in this epoch. The modern crocodilian families Crocodylidae and Alligatoridae survived the dinosaur extinction, but they are still represented by extinct genera.

Two lizard families, the anguids of the family Anguidae and the xenosaurs of the family Xenosauridae (and possibly the skink family Scincidae) are late Cretaceous carryovers, with the anguids being the most diverse group of the Paleocene. Two new families that made their appearance in the Paleocene of North America are the iguana family Iguanidae and the worm-lizard family Rhinuridae. The only Paleocene snake known is a small extinct genus of the family Boidae. This lack of snake diversity is striking.

In summary, the Paleocene herpetofauna of North America consists of nineteen living and only four extinct families. The salamanders show a striking modernization at the family level and three modern genera of salamanders occur. Amphibians are represented by seven living and two extinct families and reptiles by twelve extant and only two extinct ones. One modern turtle genus (*Trionyx*) appears for the first time. It may be said that the North American herpetofauna was quite modern at the family level by the end of the Paleocene, the first epoch of the Tertiary.

Eocene

The Eocene has yielded a moderate record of amphibians with two modern families appearing for the first time. The turtle and lizard faunas are very diverse with two families of turtles making their first appearance. Giant sea snakes are first known from the Eocene and several genera of boid snakes make their first appearance. The North American Eocene herpetofauna continues to have its major affinities with the Holarctic region and lesser affinities with tropical areas.

Five salamander families occur in the North American Eocene. The Sirenidae carries over from the late Cretaceous and the Paleocene and the salamander family Dicamptodontidae and the mole salamander family Ambystomatidae (represented by the modern genus *Ambystoma*) make their first appearance. The extinct slamander families Batrachosauroididae and Scapherpetontidae are also present. Anurans are represented by the spadefoot family Pelobatidae (including the modern genus *Scaphiopus*) and by the burrowing toad family Rhinophrynidae (represented by the modern genus *Rhinophrynus*) toward the end of the epoch. The hylid and treefrog family Hylidae (represented by the modern genus *Hyla*) appears toward the end of the epoch.

Turtle families are the same as they were in the Paleocene, except for the first appearance of the tortoise family Testudinidae and the pig-nosed softshell turtle family Carettochelyidae. The extinct turtle family Baenidae that was present in the late Cretaceous but unknown from Paleocene deposits is again present in the Eocene. Crocodilians are again represented by the modern families Crocodylidae and Alligatoridae, with the modern genus *Crocodylus* and possibly the modern genus *Alligator* first appearing in the Eocene.

The anguid family Anguidae dominates the lizard fauna of the Eocene with ten genera, some of which are very common in Eocene deposits. Iguanids, scincids, xantusiids, and varanids (present in the Cretaceous but absent in the Paleocene) are present. Snakes are represented by three families. The extinct giant sea snake family Palaeopheidae makes its first appearance, boids are represented by several newly appearing genera including some large forms, and aniliids (present in the Cretaceous but absent in the Paleocene) occur in small numbers.

In summary, the Eocene herpetofauna of North America consists of twenty-two living and five extinct families. Turtles and lizards are very diverse and there is some expansion of snakes, including the appearance of extinct giant sea snakes and several new boid genera. Amphibians are represented by six living and two extinct families and reptiles are represented by sixteen living and only three extinct families. The modern genera *Ambystoma* (Ambystomatidae), *Scaphiopus* (Pelobatidae), *Rhinophrynus* (Rhinophrynidae), *Hyla* (Hylidae), *Crocodylus,* and possibly the modern genus *Alligator* appear for the first time.

Oligocene

Oligocene amphibians are very poorly known, the turtles are modern at the family level, and the lizards, modern at the family level, are quite diverse. Several modern genera of turtles and lizards appear for the first time. Another modern salamander family shows up for the first time, and the diverse modern snake family Colubridae is represented for the first time by a single extinct genus.

The newt and salamander family Salamandridae appears for the first time in North America and is represented by the modern genus *Taricha*. Two extinct turtle families, the Baenidae and the Macrobaenidae as well as the pig-nosed softshell turtle family Carettochelyidae and the Central American river turtle family Dermatemydidae dropped out during the Eocene, leaving the Oligocene turtle fauna modern in aspect. All four Oligocene turtle families (Kinosternidae, Emydidae, Testudinidae, and Trionychidae) are present in the North American fauna today. Moreover, the recent emydid genera *Clemmys*, *Graptemys*, and *Pseudemys* (the latter two with some doubt) are reported for the first time, as is the testudinid genus *Geochelone*. Both *Crocodylus* and *Alligator*, as well as some extinct forms, are reported from the Oligocene.

Lizards are represented by eight modern families (Iguanidae, Scincidae, Xantusiidae, Xenosauridae, Anguidae, Helodermatidae, Varanidae, and Rhinuridae) and the extinct worm-lizard family Hyporhinidae. This was the last appearance of the family Varanidae in North America. Modern lizard genera that appear for the first time in the Oligocene include *Eumeces* (Scincidae), *Heloderma* (Helodermatidae), and *Rhineura* (Rhinuridae). The anguid lizards, so dominant in the Eocene, were reduced to three genera in the Oligocene. Among the snakes, the giant sea snakes of the family Palaeopheidae dropped out during the Eocene, but the Boidae was still represented by extinct genera, mainly of small size. The biggest herpetological event in the North American Oligocene may have been the first appearance of the giant modern snake family Colubridae, represented by a single genus of small snake from Colorado.

In summary, the Oligocene herpetofauna was quite modern at the family level, with eighteen modern and only one extinct family. Both families of amphibians known are extant, as are sixteen of seventeen reptile families. The family Salamandridae was reported for the first time on the basis of the modern genus *Taricha*. Gone were the extinct turtle families Baenidae and Macrobaenidae as well as the Carettochelyidae (modern southeast Asian affinities) and the Dermatemydidae (modern Central American affinities). The surviving four turtle families were all modern ones and several modern genera (or alleged modern genera) *Clemmys*, *Graptemys* (?), *Pseudemys* (?), and *Geochelone* were reported for the first time in the Oligocene. Lizards were very diverse and were represented by eight modern and one extinct family. This was the last North American appearance of the family Varanidae. Extant lizard genera reported for the first time are *Eumeces* (Scincidae), *Heloderma* (Helodermatidae), and *Rhineura* (Rhinuridae). Colubrid snakes appear for the first time on the basis of a single record of an extinct genus from Colorado.

Miocene

The North American Miocene herpetofauna was very modern at the familial level and many new modern genera appeared. The modern colubrid family Colubridae underwent a major radiation in the Miocene and poisonous snakes of the families Elapidae and Viperidae made their first appearance in North America.

Six families comprise the North American Miocene salamander fauna, the extant Cryptobranchidae, Ambystomatidae, Plethodontidae, Salamandridae, and Sirenidae, and the extinct Batrachosauroididae, which reappears in the fossil record. The lungless salamanders of the family Plethodontidae appear for the first time in the form of two modern genera, *Aneides* and *Plethodon*, and the modern genus *Notophthalmus* (Salamandridae) also makes its first appearance.

A plethora of modern anuran families make their first appearance in North

America in the Miocene. These include the Bufonidae (bufonids and toads), Ranidae (ranids), Pelodytidae (parsley frogs), Microhylidae (microhylids), and Leptodactylidae (leptodactylids). The Pelobatidae and Hylidae are also found in the North American Miocene. Modern genera showing up for the first time include *Bufo* (Bufonidae), *Rana* (Ranidae), *Gastrophryne* (Microhylidae), *Leptodactylus* (Leptodactylidae), and *Acris* and *Pseudacris* (Hylidae).

Turtles are represented by six families (Chelydridae, Emydidae, Testudinidae, Kinosternidae, Trionychidae, and Pelomedusidae), with the pelomedusids reappearing in the North American fossil record for the first time since the late Cretaceous. Modern turtle genera showing up for the first time are *Chelydra* and *Macroclemys* (Chelydridae), *Chrysemys*, *Emydoidea,* and *Terrapene* (Emydidae), and *Kinosternon* and *Sternotherus* (Kinosternidae). The crocodilians *Crocodylus* and *Alligator* are present, as are two extinct genera.

Six families of lizards occur in the North American Miocene including the Iguanidae, Scincidae, Teiidae, Anguidae, Helodermatidae, and Rhinuridae. Modern lizard genera showing up for the first time are *Sceloporus*, *Holbrookia*, *Anolis*, *Leiocephalus*, and *Dipsosaurus* (Iguanidae), *Cnemidophorus* (Teiidae), and *Ophisaurus*, *Gerrhonotus*, and *Anniella* (Anguidae).

Snakes are represented by five families, including the Aniliidae, Boidae, Colubridae, Elapidae, and Viperidae. The elapids and the viperids appear for the first time in North America in the Miocene. The giant modern snake family Colubridae underwent a major radiation in the Miocene that produced eleven extinct colubrid genera and eight modern ones. This was the largest radiation of any reptilian group in the Tertiary, even surpassing the anguid lizard radiation in the Eocene! Modern snake genera showing up for the first time include *Charina* (Boidae) and *Heterodon*, *Diadophis*, *Coluber*, *Elaphe*, *Lampropeltis*, *Salvadora*, *Nerodia*, and *Thamnophis* (Colubridae).

In summary, the herpetofauna of North America became essentially modern in the Miocene. Of thirty-one families recorded only one was extinct. Of the thirteen amphibian families recorded, only one is extinct and none of the nineteen reptile families recorded are extinct! The following modern herpetological families appeared for the first time: salmanders—Plethodontidae; anurans—Bufonidae, Ranidae, Pelodytidae, and Microhylidae; and snakes—Elapidae and Viperidae. Three modern genera of salamanders, six modern genera of anurans, seven modern genera of lizards, and nine modern genera of snakes made their first appearance in the Miocene.

Pliocene

The Pliocene is much the shortest epoch of the Tertiary, lasting only about 3 million years. The Pliocene essentially represents only one Land Mammal Age, the Blancan. Herpetological sites are rare in the North American Pliocene, but those that are present indicate a very modern herpetofauna (Estes and Baez, 1985). The only extinct genera left are the crocodilian *Gavialosuchus* (a somewhat gaviallike form) and the extinct boid genus *Ogmophis,* which makes its last appearance in Texas.

The lizard fauna is dominated by iguanids, and the xantusiid genus *Xantusia* appears for the first time. Snakes are the most commonly recorded reptiles of the Pliocene and most of them are colubrids. Gone are all the extinct "archaic" colubrid genera of the Miocene, but at least eight modern genera are present, including the first appearance of *Rhinocheilus*. All three modern genera of the poisonous snake family Viperidae, *Agkistrodon*, *Sistrurus*, and *Crotalus,* are present in the Pliocene. Several of these snake genera have been assigned to modern species, and it is quite likely that many modern species of amphibians and reptiles originated in the Pliocene, setting the stage for an essentially modern herpetofauna at the specific level at the onset of the Pleistocene.

Summary

The North American herpetofauna from the late Cretaceous through the Pliocene had its major affinities with the Holarctic region, with the Bering land bridge between northern Eurasia and North America being the most important herpetofaunal continental link. The North American herpetofauna was already surprisingly modern at the familial level by late Cretaceous times, with modern families outnumbering extinct families by about three to one. Salamanders, anurans, turtles, crocodilians, and lizards were common, but snakes were rare and represented by only two families. A few living genera have been reported from the late Cretaceous, but these have been seriously questioned.

Many families of salamanders, anurans, turtles, crocodilians, and lizards survived the dinosaur extinction episode at the end of the Cretaceous and made it into the Paleocene, and several modern families made their appearance in that epoch. Unquestioned modern genera of the Paleocene include the salamanders *Cryptobranchus* and *Necturus* and the turtle *Trionyx*. Living families outnumbered extinct families by about four to one in the Paleocene.

The Eocene produced a good record of amphibians and turtles, and lizards flourished. Additional modern genera began to appear and modern families still outnumbered extinct families by about four to one. The most diverse family of the Eocene herpetofauna was the lizard family Anguidae, which had a major radiation that produced at least ten Eocene genera. The Eocene was the epoch of the giant sea snakes in estuarine deposits in North America, and this group became extinct by the end of the epoch. Terrestrial snakes were still insignificant.

The Oligocene saw almost the complete modernization of the North American herpetofauna at the family level as modern families now outnumbered extinct families by an overwhelming eighteen to one! A very important event was the arrival from Asia of the first snake of the giant modern snake family Colubridae, represented by a single genus and species from Colorado.

The Miocene saw the further modernization of the North American herpetofauna at the generic level. The Miocene was the time of an explosive radiation of the family Colubridae. This family went from one genus in the Oligocene to nineteen or more genera by the end of the Miocene, and this surpassed even the large radiation of the anguid lizards in the Eocene. The Pliocene is poorly represented by amphibians and reptiles in North America, but modern species of snakes have been recorded; and it is believed that many modern species originated in the epoch, thus setting the herpetological stage for the Pleistocene.

4

A BESTIARY

This chapter consists of synoptic remarks about all the Pleistocene orders, families, and genera of amphibians and reptiles from mainland North America (excluding Mexico) with references to regions delineated in this book in which their species occur. The precise localities of these Pleistocene species will be found in the regional accounts in Chapter 5.

The arrangement of taxonomic groups and nomenclature mainly follows the widely distributed checklist of Banks, McDiarmid, and Gardner (1987). Exceptions include changes in the spelling of several scientific names since the 1987 publication (e.g., Conant and Collins, 1991); the fact that snakes, lizards, and amphisbaenians are considered orders rather than suborders following Estes (1983) and Rage (1984); and the fact that the Rhinuridae is considered a separate family (Estes, 1983).

Class Amphibia Linnaeus, 1758: Amphibians

Order Caudata Oppel, 1811: Salamanders

Eight families of salamanders occur natively in the United States and Canada today and seven of them are represented by Pleistocene fossils. The unrepresented family, the Dicamptodontidae (giant and Olympic salamanders), occurs in the northern Pacific area of the United States and southern tip of British Columbia. It is represented by two genera and four species. The seven families represented by North American Pleistocene fossils are:

Cryptobranchidae (hellbenders and giant salamanders)
Sirenidae (sirens)
Salamandridae (newts and salamanders)
Proteidae (olms, mudpuppies, and waterdogs)
Amphiumidae (amphiumas)

Ambystomatidae (mole salamanders)
Plethodontidae (lungless salamanders)

Fourteen genera and twenty-three species of salamanders have been reported as North American Pleistocene fossils. None of these genera are extinct, but two extinct species have been described.

Family Cryptobranchidae Fitzinger, 1826

Hellbenders and Giant Salamanders (Figs. 3A, 16A)

This family is represented in North America today by a single monotypic genus, *Cryptobranchus*, a very large, brownish or brownish red, irregularly shaped salamander called the hellbender. Hellbenders are permanently aquatic forms with external gills and external fertilization that prefer clear streams and rivers, mainly in the Appalachian and Ozark regions of the United States. Osteological characters include the presence of angular and maxillary bones but the absence of septomaxillary and lacrimal bones. The columella is free from the otic capsule. The vertebrae (Fig. 16A) are amphicoelous and simple in structure, with a low, indistinct neural spine, simple rib-bearers, and a relatively wide hemal keel.

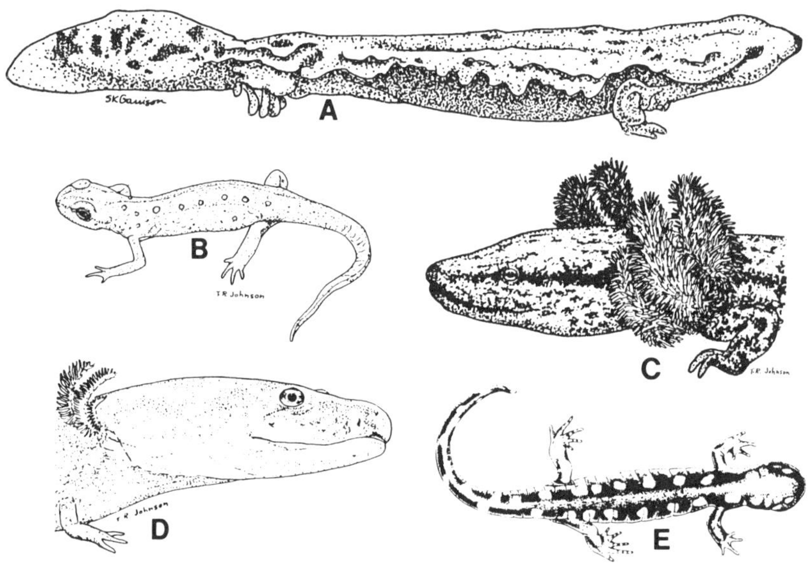

FIGURE 3. Representatives of North American salamander families. (A) *Cryptobranchus alleganiensis* (Cryptobranchidae). (B) *Notophthalmus viridescens* (Salamandridae). (C) *Necturus maculosus* (Proteidae). (D) *Amphiuma means* (Amphiumidae). (E) *Ambystoma opacum* (Ambystomatidae). These species occurred in the Pleistocene of North America. [Drawings by S. K. Garrison (A) and T. R. Johnson (B–E), permission of *Herpetological Review*]

Cryptobranchus Leuckart, 1821: Hellbender

Only one modern species, *Cryptobranchus alleganiensis* (Fig. 3A), occurs in North America today, and it also occurs as a North American Pleistocene fossil. An extinct Pleistocene species, *Cryptobranchus guildayi* (Fig. 16A), has also been described and it will be addressed in the appropriate regional account.

Regional Pleistocene Occurrences

Cryptobranchus alleganiensis. Hellbender: Appalachian region.
Cryptobranchus guildayi. Guilday's hellbender: Appalachian region.

Additionally, *Cryptobranchus* sp. has been reported from the Ozark region.

Family Sirenidae Gray, 1825

Sirens (Figs. 17B, 33A,B)

This family is restricted to North America today and is represented by two genera, *Pseudobranchus* and *Siren,* which occur mainly in the southeastern United States. These genera are eellike, permanently aquatic forms with external gills and with only one pair of obsolete forelimbs. They employ external fertilization. Both genera are known as Pleistocene fossils and no extinct Pleistocene genera have been described.

Osteological characters include the absence of the angular, septomaxillary, and lacrimal bones. The maxillary bone may be either present or absent. The columella is free from the otic capsule. The vertebrae (Figs. 17B, 33A,B) are amphicoelous and have a complex structure, with a high neural spine with a V-shaped posterior expansion, as well as elaborate V-shaped structures on the lateral portions of the neural arches and a very sharp hemal keel.

Pseudobranchus Gray, 1825: Dwarf Sirens

Dwarf sirens are very small, yellowish brown, eellike salamanders. They inhabit marshes and swamps. A single modern species, *Pseudobranchus striatus*, occurs in the southeastern United States. This species is unknown as a Pleistocene fossil, but an extinct species, *Pseudobranchus robustus* (Fig. 33A), has been described. This extinct species will be addressed in the appropriate regional account.

Regional Pleistocene Occurrences

Pseudobranchus robustus. Robust dwarf siren: Southeastern region.

Siren Linnaeus, 1766: Sirens

Sirens are medium-sized or large, brownish, eellike salamanders that prefer marshy or swampy habitats. They occur today in the southeastern and south-central United States. Two modern species are known, but only one, *Siren lacertina*, a large form of the southeastern United States, is known as a Pleistocene fossil. No extinct Pleistocene species of *Siren* have been described.

Regional Pleistocene Occurrences

Siren lacertina. Greater siren: Southeastern region.

Family Salamandridae Gray, 1825

Newts and Salamanders (Figs. 3B, 17C)

This family is represented in North America today by only two genera, *Notophthalmus* and *Taricha*, and only the former is known as a North American Pleistocene fossil. No extinct Pleistocene genera have been described. Salamandrids often have an aquatic (newt) and a warty terrestrial (eft) stage in their life cycle, but the patterns of occurrence of these stages often differ in different species and populations. Salamandrids lack external gills in the adult stages and have well-developed limbs and internal fertilization.

Osteological characters include the absence of the angular, septomaxillary, and lacrimal bones. The maxilla is present. The columella is fused to the otic capsule. Vertebrae may be simple with low neural spines (*Salamandra* and some *Triturus* of the Old World), or rather complex (Fig. 17C), with high neural spines, often capped with dermal encrustations (*Notophthalmus*). Anterior bony caps often fuse to the centrum of salamandrid vertebrae, creating a secondary opisthocoelous condition.

Notophthalmus Rafinesque, 1820: Eastern Newts

Eastern newts (Fig. 3B) are small salamanders with a brown, reddish or yellowish spotted newt stage and a bright red warty eft stage. They prefer damp habitats near or in standing water. These salamanders occur in about the eastern one-third of North America including southeastern Canada today. Three species of eastern newts occur in North America today, but only one, *Notophthalmus viridescens*, has been recorded as a Pleistocene fossil. No extinct Pleistocene species have been described.

Regional Pleistocene Occurrences

Notophthalmus viridescens. Eastern newt: Appalachian and Southeastern regions.

In addition, *Notophthalmus* sp. has been identified as a fossil from the Southern Plains region.

Family Proteidae Gray, 1825

Olms, Mudpuppies, and Waterdogs (Figs. 3C, 17D)

This family is represented in North America today by a single genus, *Necturus* (Fig. 3C), a large, brownish or reddish brown, darkly spotted, flattened salamander. These animals are permanently aquatic with external gills, but they have internal fertilization. They occur in a wide variety of aquatic habitats. This genus occurs widely in about the eastern one-third of the United States and in southeastern Canada.

Osteological characters include the absence of angular, septomaxillary, lacrimal, and maxillary bones. The columella is free from the otic capsule. The vertebrae (Fig. 17D) are amphicoelous and simple in structure, with a low, indistinct neural spine, simple rib-bearers, and a moderately sharp hemal keel.

Necturus Rafinesque, 1819: Mudpuppies and Waterdogs

Five modern North American species occur in this genus, but only one, *Necturus maculosus* (Fig. 3C) is known as a Pleistocene fossil. No extinct Pleistocene species have been described.

Regional Pleistocene Occurrences

Necturus maculosus. Mudpuppy: Appalachian and Ozark regions.

Additionally, *Nectrurus* sp. has been reported outside of its modern range in Florida in the Southeastern region.

Family Amphiumidae Gray, 1825

Amphiumas (Fig. 3D)

Amphiumas are restricted to North America, where today they are represented by a single genus, *Amphiuma* (Fig. 3D), a very large, blackish, eellike salamander that prefers shallow, still aquatic situations. These are permanently aquatic animals, with two pairs of tiny limbs, but they lack external gills in the adult form. They have internal fertilization. Amphiumas occur in the southeastern United States west to Texas, mainly in coastal plain areas.

Osteological characters include the absence of the angular, septomaxillary, and lacrimal bones. The maxillary bone is present. The columella is fused to the otic capsule. The vertebrae are amphicoelous and complex, with a high, thin, undivided neural spine. The rib-bearers are fused and expanded. The hemal keel is very thin, but well-developed.

Amphiuma Garden, 1821: Amphiumas

Three modern North American species of *Amphiuma* are currently recognized, but only one, *Amphiuma means* (Fig. 3D), a large form of the southeastern United States, is known as a Pleistocene fossil. No extinct Pleistocene species have been described.

Regional Pleistocene Occurrences

Amphiuma means. Two-toed amphiuma: Southeastern and Southern Plains regions.

Family Ambystomatidae Hallowell, 1856

Mole Salamanders (Figs. 3E, 15, 17E)

Mole salamanders are represented in North America by two genera, the Mexican *Rhyacosiredon* and *Ambystoma* (Fig. 3E), which contain medium-sized salamanders that may be spotted, barred, flecked with blue or white, or rather plain-colored. These animals are mainly terrestrial fossorial forms in the adult stage, but species in both genera may exist as neotenic aquatic forms. The limbs are well-developed and fertilization is internal. These animals occur very widely throughout temperate North America.

Osteological characters include the absence of angular and lacrimal bones and the presence of septomaxillary and maxillary bones. The columella is fused to the otic capsule. The vertebrae (Figs. 15, 17E) are amphicoelous and simple. The neural spines are indistinct and the rib-bearers are simple and partially divided. The bottom of the centrum is smooth and unkeeled.

Ambystoma Tschudi, 1838: Mole Salamanders

At least sixteen modern species of mole salamanders occur in North America today and some of these exist as all-female populations. At least six of these species have been recorded as Pleistocene fossils. No extinct Pleistocene species have been described, but one Irvingtonian species from Colorado awaits description.

Regional Pleistocene Occurrences

Ambystoma laterale. Blue-spotted salamander: Northeastern/Great Lakes and Northern Plains regions.

Ambystoma jeffersonianum complex. Jefferson's salamander or one of its close relatives: Appalachian region.

Ambystoma opacum. Marbled salamander: Appalachian and Ozark regions.

Ambystoma maculatum. Spotted salamander: Appalachian, Ozark, and Northern Plains regions.

Ambystoma texanum. Small-mouthed salamander: Ozark and Southern Plains regions.

Ambystoma tigrinum. Tiger salamander: Appalachian, Southeastern, Ozark, Northern Plains, Southern Plains, and Rocky Mountain/Great Basin regions.

Family Plethodontidae Gray, 1850

Lungless Salamanders (Figs. 4, 17F)

The lungless salamanders are the most widespread and diverse family of caudates in North America today, with sixteen genera and at least seventy-eight species. They occur throughout much of temperate North America but are most diverse in the Appalachian region of the United States, especially the southern Appalachians.

Seven of these genera are known as Pleistocene fossils and no extinct Pleistocene genera have been described. These animals occur in many damp terrestrial habitats, as well as in or near streams or in caves. Some are neotenic cave species (Fig. 4E,F). Plethodontids are the only salamanders that are lungless and have a nasolabial groove. All of them have internal fertilization.

Osteological characters include the absence of angular and lacrimal bones and the presence of septomaxillary and maxillary bones. The columella is fused to the otic capsule. The vertebrae are variable, but all have divided rib-bearers. Most of them have moderately well-developed neural spines, but they are not as high as in the Sirenidae and Amphiumidae. Some species (Fig. 17F) have an ossified cap that fits into and sometimes fuses with the anterior cotyle of the centrum, creating a falsely opisthocoelous condition.

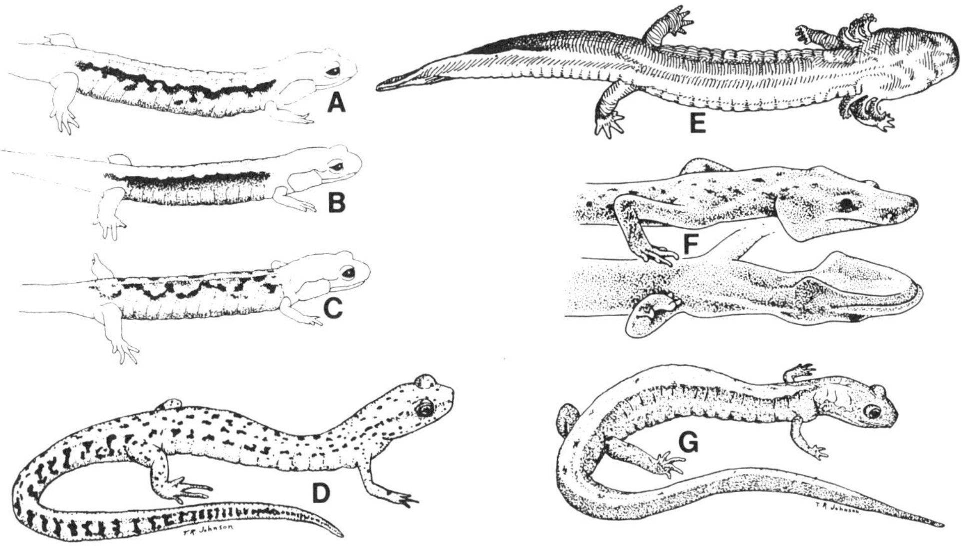

FIGURE 4. Some North American salamanders of the family Plethodontidae. (A–C) *Desmognathus ochrophaeus* pattern variations. (D) *Eurycea longicauda*. (E) Neotenic, and (F) metamorphosed *Gyrinophilus palleucus*, a normally neotenic cave-dwelling species. (G) *Plethodon cinereus*. All these genera and the species and *Desmognathus ochrophaeus* occurred in the Pleistocene of North America. [A–C from Tilley, 1969; E and F from McCrady, 1954, permission of *Copeia;* D and G, drawings by T. R. Johnson, permission of *Herpetological Review*]

Aneides Baird, 1849: Climbing Salamanders

Climbing salamanders are small brown or black salamanders that are often cryptically colored with lichenlike spots. One species occurs in the southern Appalachians and the others in New Mexico, the Pacific Coast states, and Vancouver Island. Only one of the five species, the western *Aneides lugubris*, has been recorded as a Pleistocene fossil. No extinct Pleistocene species have been described.

Regional Pleistocene Occurrences

Aneides lugubris. Arboreal salamander: Pacific Coast region.

Desmognathus Baird, 1850: Dusky and Seal Salamanders

These medium-sized salamanders are mainly dusky or dull-colored, although some may be striped or lightly speckled (Fig. 4A–C). Most of these animals live near or in small streams. They mainly occur in the southeastern United Sates, but some range into the northeastern United States and southeastern Canada. Three of the eleven modern species have been identified as Pleistocene fossils. No extinct Pleistocene species have been described.

Regional Pleistocene Occurrences

Desmognathus fuscus. Dusky salamander: Appalachian region.

Desmognathus monticola. Seal salamander: Appalachian region.
Desmognathus ochrophaeus. Mountain dusky salamander: Appalachian region.

Eurycea Rafinesque, 1822: Brook Salamanders

Brook salamanders (Fig. 4D) are small to medium sized, brightly colored yellowish, orangish, or redddish animals with dark stripes or spots. Most of them occur near small streams, but some live in caves, and others are neotenic cave forms. Brook salamanders are widely distributed in the eastern and central parts of the United States and in eastern Canada today. Two of the eight modern *Eurycea* species have been reported as Pleistocene fossils, but no extinct Pleistocene species have been described.

Regional Pleistocene Occurrences

Eurycea bislineata. Two-lined salamander: Appalachian region.
Eurycea lucifuga. Cave salamander: Appalachian region.

Additionally, *Eurycea* sp. has been reported from the Pleistocene of the Ozark region.

Gyrinophilus Cope, 1869: Cave and Spring Salamanders

These rather large salamanders have a yellowish body color with brownish reticulations or spots. They are either cave- (Fig. 4E,F) or spring-dwelling forms. This genus mainly occurs in the Appalachian region of the United States, but one form reaches extreme southeastern Canada. There are at least two modern species, only one of which, *Gyrinophilus porphyriticus*, has been reported as a Pleistocene fossil. No extinct Pleistocene species have been described.

Regional Pleistocene Occurrences

Gyrinophilus porphyriticus. Spring salamander: Appalachian region.

Additionally, *Gyrinophilus* sp. has been reported from the Appalachian region, where it was suggested that the fossil remains might represent a neotenic cave-dweller.

Hydromantes Gistel, 1848: Web-Toed Salamanders

These small, brownish to grayish, web-toed salamanders are excellent climbers. The genus is restricted to the rocky uplands of central California today. Three modern species have been described, but only "*Hydromantes* sp." has been reported as a Pleistocene fossil.

Regional Pleistocene Occurrences

Hydromantes sp. Web-toed salamander: Pacific Coast region.

Plethodon Tschudi, 1838: Woodland Salamanders

These small (Fig. 4G) to medium-sized salamanders range from dull to brightly colored species. Some are shiny black with tiny white spots, others are rather bright red, while still others have bright red cheek or leg patches. These animals prefer woodland habitats and are widely distributed throughout temperate North

America. At least twenty-seven species have been described in the modern fauna, but only two of these species have been recorded as Pleistocene fossils. No extinct Pleistocene species have been described.

Regional Pleistocene Occurrences

Plethodon glutinosus. Slimy salamander: Appalachian, Southeastern, Ozark, and Southern Plains regions.

Plethodon dorsalis. Zig-zag salamander: Ozark region.

Pseudotriton Tschudi, 1838: Red and Mud Salamanders

These are rather large reddish or bright red salamanders with small dark spots. They mainly occur in moist woodland habitats. These salamanders occur in the eastern United States, mainly in the southeastern portion. Only one of the two modern species, *Pseudotriton ruber*, has been identified as a Pleistocene fossil. No extinct Pleistocene species have been described.

Regional Pleistocene Occurrences

Pseudotriton ruber. Red salamander: Appalachian region.

Order Anura Rafinesque, 1815: Frogs and Toads

Seven families of anurans occur natively in the United States and Canada today and six of them are recognized as Pleistocene fossils. The family that is not represented is the Leiopelmatidae (tailed frogs), which consists of a single species (*Ascaphus truei*) that occurs in extreme northwestern United States and southwestern Canada. The six families that are represented by Pleistocene fossils in North America are:

Pelobatidae (pelobatids and spadefoots)
Leptodactylidae (leptodactylids)
Bufonidae (bufonids and toads)
Hylidae (hylids and treefrogs)
Microhylidae (microhylids)
Ranidae (ranids)

Of the seven genera and forty-six species of anurans of the six families that occur as North American Pleistocene fossils, no genera, two species, and one subspecies are currently recognized as extinct. These will be discussed in the appropriate regional accounts.

Family Pelobatidae Bonaparte, 1850

Pelobatids and Spadefoots (Figs. 5A, 19A, 21C)

This family is represented in North America by a single genus, *Scaphiopus* (Fig. 5A), whose species are medium-sized, rather plain, blotched or lightly striped

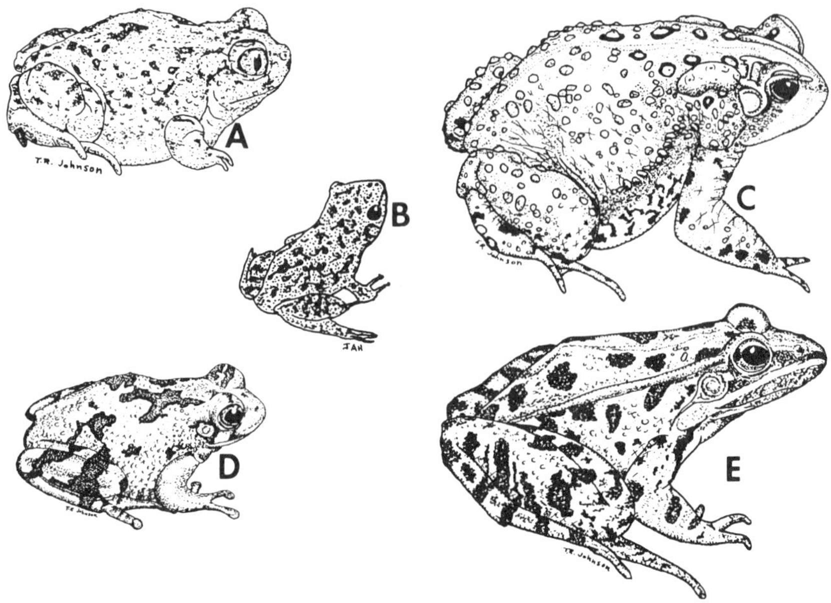

FIGURE 5. Representatives of North American anuran families. (A) *Scaphiopus couchii* (Pelobatidae). (B) *Syrrhophus marnockii* (Leptodactylidae). (C) *Bufo americanus* (Bufonidae). (D) *Hyla versicolor* (Hylidae). (E) *Rana utricularia* (Ranidae). All of these species occurred in the Pleistocene of North America. [A and C–E, drawings by T. R. Johnson, permission of *Herpetological Review;* B, sketch by the author]

toadlike forms with large eyes and a cornified spade on the hind foot. All modern species are burrowing forms that may remain underground for extended periods of time. They breed infrequently but explosively, and the larvae usually metamorphose in only a few days. Modern species of the genus range mainly throughout the xerophytic areas of the central and western United States and in central-southern Canada. No extinct Pleistocene genera of pelobatids have been described.

Osteological characters include an arciferal pectoral girdle and vertebrae that have procoelous centra throughout the vertebral column. The sacral diapophyses (Fig. 21C) are greatly expanded and are fused to the urostyle. The ilium (Fig. 19A) lacks a dorsal blade and the dorsal prominence is usually obsolete.

Scaphiopus Holbrook, 1836: Spadefoots

Five of the six modern species are known as Pleistocene fossils. No extinct Pleistocene species have been described.

Regional Pleistocene Occurrences

Scaphiopus holbrookii. Eastern spadefoot: Appalachian, Southeastern, and Ozark regions.

Scaphiopus bombifrons. Plains spadefoot: Northern Plains and Rocky Mountain/Great Basin regions.

Scaphiopus couchii. Couch's spadefoot: Southern Plains and Rocky Mountain/Great Basin regions.
Scaphiopus intermontanus. Great Basin spadefoot: Rocky Mountain/Great Basin region.
Scaphiopus hammondii. Western spadefoot: Rocky Mountain/Great Basin region.

Family Leptodactylidae Werner, 1896

Leptodactylids (Figs. 5B, 19B)

This family is represented in North America by four genera that occur mainly in Texas and New Mexico, with one form reaching southern Arizona. The modern North American leptodactylid fauna is diminished compared to that of the tropical parts of the New World, where hundreds of species have been described. Only two leptodactylid genera are represented in North America as Pleistocene fossils. No extinct Pleistocene genera have been described.

Osteological characters include a pectoral girdle that is mainly arciferal and vertebrae that are procoelous throughout the vertebral column. The sacral diapophyses are rounded and the sacrum is free from the urostyle. The ilium (Fig. 19B) often has an ilial blade, but it is usually not as well developed as in the Ranidae.

Hylactophryne Lynch, 1968: Barking Frogs

Barking frogs are medium-sized, brownish, darkly blotched, toadlike, but wartless anurans that are often found in caves. They occur in Texas, New Mexico, and Arizona today, and they are represented by a single species, *Hylactophryne augusti*. This species is represented by Pleistocene fossils, but no extinct Pleistocene species have been described.

Regional Pleistocene Occurrences
Hylactophryne augusti. Barking frog: Southern Plains region.

Syrrhophus Cope, 1878: Chirping Frogs

Chirping frogs are small, greenish, brownish-mottled frogs that, in the area of this book, are confined to Texas. Three Texas species occur, one of which, *Syrrhophus marnockii* (Fig. 5B, 19B), has been reported as a Pleistocene fossil. This species occurs today mainly in the Balcones escarpment area of Texas, where it may occur in caves. No extinct Pleistocene species of chirping frogs have been described.

Regional Pleistocene Occurrences
Syrrophus marnockii. Cliff chirping frog: Southern Plains region.

Family Bufonidae Gray, 1825

Bufonids and Toads (Figs. 5C, 19C,D, 21A, 34A,B, 37)

This almost cosmopolitan family is represented in North America by a single genus, *Bufo* (Fig. 5C), which is represented by medium- to large-sized, short-legged, plump anurans with a warty skin and prominent paratoid glands near the head, which secrete a distasteful or poisonous bufotoxin. The modern North American species are ubiquitous in their choice of habitat and frequently occur long distances from a permanent water source. All of them, however, go through a tadpole stage. This genus occurs throughout the United States and in all but the coldest areas of Canada. No extinct Pleistocene bufonid genera have been described.

Osteological characters include an arciferal pectoral girdle and a vertebral column that has procoelous vertebrae throughout its extent. The sacral diapophyses may be cylindrical or dilated (Fig. 21A) and the sacrum is not fused to the urostyle. An ilial blade is absent, but an ilial prominence arises dorsally (Fig. 19C,D). This prominence may be irregular in shape to highly defined and ranges in height from low to high.

Bufo Laurenti, 1768: Toads

Nine of the eighteen modern North American species are known as Pleistocene fossils, and one extinct species, *Bufo repentinus* (Fig. 34A), and one extinct subspecies, *Bufo woodhousii bexarensis* (Fig. 37), have been described. These forms will be addressed under the appropriate regional accounts.

Regional Pleistocene Occurrences

Bufo americanus. American toad: Northeastern/Great Lakes, Appalachian, Ozark, and Northern Plains regions.

Bufo terrestris. Southern toad: Appalachian and Southeastern regions.

Bufo quercicus. Oak toad: Southeastern region.

Bufo woodhousii. Woodhouse's toad: Appalachian, Southeastern, Ozark, Northern Plains, Southern Plains, and Rocky Mountain/Great Basin regions.

Bufo woodhousii bexarensis. Bexar County, Texas, toad: Southern Plains region.

Bufo hemiophrys. Canadian toad: Northern Plains and (?) Southern Plains regions.

Bufo repentinus. Extinct Plains toad: Northern Plains region.

Bufo cognatus. Great Plains toad: Northern Plains, Southern Plains, and Rocky Mountain/Great Basin regions.

Bufo valliceps. Gulf Coast toad: Southern Plains region.

Bufo boreas. Western toad: Rocky Mountain/Great Basin and Pacific Coast regions.

Bufo punctatus. Red-spotted toad: Rocky Mountain/Great Basin and Pacific Coast regions.

Family Hylidae Gray, 1825

Hylids and Treefrogs (Figs. 5D, 19E,F, 32, 35)

This mainly tropicolitan family is represented in North America by seven genera. The family occurs throughout most of the United States and much of Canada except for the far western and very coldest areas. North American hylids range from arboreal species to species of high marsh grasses and mud flats. Some have well-developed toe pads, whereas others have rudimentary ones. Some of them are bright green in color and have very attractive stripes or spots. Some can change color from gray to bright green. All North American species have a larval stage so that aquatic situations for breeding are essential. Three of the seven genera that occur in North America today are represented by Pleistocene fossils. No extinct Pleistocene genera have been described, but two extinct Pleistocene species have been described and will be discussed in the appropriate regional account.

Osteological characters include an arciferal pectoral girdle and a vertebral column that is procoelous throughout its extent. The sacral diapophyses are usually dilated by may occasionally be round. The sacrum is not fused to the coccyx. The hylid ilium (Figs. 19E,F, 32, 35) typically has a small, usually rounded, dorsal protuberance that is directed laterally. It lacks a dorsal blade, but a dorsal ridge may be present. Intercalary cartilages or bones are present in the digits.

Acris Duméril and Bibron, 1841: Cricket Frogs

These are small, brownish green, warty little frogs with rudimentary toe pads. They are mainly to be found in freshwater sand or mudflat habitats today. This genus occurs widely in temperate regions of the eastern and central United States. Two species occur in the modern fauna, both of which have been reported as Pleistocene fossils. No extinct Pleistocene species have been described.

Regional Pleistocene Occurrences

Acris crepitans. Northern cricket frog: Northern Plains and Southern Plains regions.

Acris gryllus. Southern cricket frog: Southeastern region.

Hyla Laurenti, 1768: Treefrogs

Species of this genus are small to medium-sized (Fig. 5D), usually green, gray, or brownish frogs that may be brilliantly striped or spotted. They are arboreal frogs with long legs and well-developed toe pads. This genus is widespread in North America today in temperate areas, but it is most diverse in the southeastern United States. At least six of the thirteen modern North American species are represented by Pleistocene fossils, and two extinct species, *Hyla baderi* (Fig. 32) and the questionable *Hyla holmani* (Fig. 35), have been described. These extinct species will be addressed in the appropriate regional accounts.

Regional Pleistocene Occurrences

Hyla versicolor. Gray treefrog: Appalachian, Northern Plains and Southern Plains regions. (The cryptic counterpart of *Hyla versicolor*, *Hyla chrysoscelis* [Cope's gray treefrog], may also be represented by some of this fossil material.)

Hyla gratiosa. Barking treefrog: Appalachian and Southeastern regions.

Hyla baderi. Bader's treefrog: Southeastern region.
Hyla cinerea. Green treefrog: Southeastern region.
Hyla femoralis. Pine woods treefrog: Southeastern region.
Hyla holmani. Holman's treefrog: Southern Plains region.
Hyla squirella. Squirrel treefrog: Southeastern region.
Hyla arenicolor. Canyon treefrog: Rocky Mountain/Great Basin region.

Pseudacris Fitzinger, 1849: Chorus Frogs and Peepers

These are very small, reddish, brownish, or grayish striped or spotted frogs that have rudimentary toe pads. They occur in marshes or woodlands in much of North America except for very dry areas and the Pacific coastal states, and they extend far northward in Canada. All of them are most obvious in the spring breeding season. Nine modern North American species are recognized, seven of which are represented by Pleistocene fossils. No extinct Pleistocene species have been described.

Regional Pleistocene Occurrences

Pseudacris crucifer. Spring peeper: Appalachian region.
Pseudacris triseriata. Striped chorus frog: Appalachian, Northern Plains, Southern Plains, and Rocky Mountain/Great Basin regions.
Pseudacris nigrita. Southern chorus frog: Southeastern region.
Pseudacris ornata. Ornate chorus frog: Southeastern region.
Pseudacris clarkii. Spotted chorus frog: Northern Plains and Southern Plains regions.
Pseudacris streckeri. Strecker's chorus frog: Southern Plains region.
Pseudacris regilla. Pacific chorus frog: Pacific Coast region.

Family Microhylidae Gunther, 1859

Microhylids (Fig. 20F,G,H)

North American frogs of this widespread family are small, brownish, fat-bodied forms with a narrow snout. They live in moist habitats and feed principally on ants. Microhylids occur in the southeastern United States and in the southern Great Plains today. They consist of two genera, *Gastrophryne,* which has been recorded as a Pleistocene fossil, and *Hypopachus,* and no extinct Pleistocene genera have been described.

Osteological characters include a firmisternal pectoral girdle and a vertebral column that is procoelous throughout the vertebral column in most genera, except in some, where the sacral vertebra is opisthocoelous. The microhylid sacral diapophysis is expanded. The ilium (Fig. 20F,G,H) lacks a dorsal blade and has a very characteristic triangular dorsal prominence.

Gastrophryne Fitzinger, 1843: Narrow-Mouthed Toads

This genus of small, brownish, plump, narrow-snouted frog prefers damp terrestrial habitats. These animals occur in the southern parts of North America, in the eastern and central plains regions. Two modern species are recognized, both of which are represented by Pleistocene fossils. No extinct Pleistocene species have been described.

Regional Pleistocene Occurrences

Gastrophryne carolinensis. Eastern narrow-mouthed toad: Appalachian and Southeastern regions.

Gastrophryne olivacea. Western narrow-mouthed toad: Southern Plains region.

Family Ranidae Gray, 1825

Ranids (Figs. 5E, 18, 20A-E, 21B)

This very large, almost cosmopolitan family is represented in North America by a single large genus, *Rana* (Fig. 5E), which consists of medium-sized to large, greenish or greenish brown, rather plain, striped or spotted frogs with pointed snouts and webbed feet. They are mainly aquatic frogs that prefer quiet water, but some species wander far into grassy or damp woodland areas. They all have an aquatic larval stage. Ranid frogs occur throughout most of the United States and far northward into Canada. This genus is very widespread as a Pleistocene fossil, but no extinct Pleistocene genera have been described.

Osteological characters include a pectoral girdle that is firmisternal. The vertebrae are all procoelous except for the sacral vertebra, which is opisthocoelous. The sacrum (Fig. 21B) has cylindrical diapophyses. The *Rana* ilium (Figs. 18, 20A–E) has a well-developed ilial blade.

Rana Linnaeus, 1758: True Frogs

Twenty-five species of this genus occur in North America today, at least ten of which are known as Pleistocene fossils. No extinct Pleistocene species have been described.

Regional Pleistocene Occurrences

Rana catesbeiana. Bullfrog: Northeastern/Great Lakes, Appalachian, Southeastern, Ozark, Northern Plains, Southern Plains, and Rocky Mountain/Great Basin regions.

Rana clamitans. Green frog: Northeastern/Great Lakes, Appalachian, and Northern Plains regions.

Rana pipiens. Northern leopard frog: Northeastern/Great Lakes and Appalachian regions.

Rana pipiens complex. Northern leopard frog or one of its close relatives: Northeastern/Great Lakes, Appalachian, Ozark, Northern Plains, Southern Plains, and Rocky Mountain/Great Basin regions.

Rana sylvatica. Wood frog: Northeastern/Great Lakes, Appalachian, and Northern Plains regions.

Rana palustris. Pickerel frog: Appalachian and Ozark regions.

Rana utricularia. Southern leopard frog: Appalachian and Southeastern regions.

Rana areolata. Crawfish frog: Southeastern region.

Rana grylio. Pig frog: Southeastern region.

Rana blairi. Plains leopard frog: Northern Plains region.
Rana aurora. Red-legged frog: Pacific Coast region.

Class Reptilia Laurenti, 1768: Reptiles

Order Testudines Batsch, 1788: Turtles and Tortoises

Five families of nonmarine turtles and tortoises occur natively in the United States and Canada today, and all of them are represented by Pleistocene fossils. These five families are:

Kinosternidae (mud and musk turtles)
Chelydridae (snapping turtles)
Emydidae (emydid turtles)
Testudinidae (tortoises)
Trionychidae (softshell turtles)

Fifteen genera and thirty-three species of these families occur as North American Pleistocene fossils. None of these genera are extinct, but at least seven extinct species and three subspecies are currently recognized. These extinct forms will be discussed under the appropriate regional accounts.

Family Kinosternidae Baur, 1893

Mud and Musk Turtles (Figs. 6A, 23A)

This family occurs in North America mainly in the temperate regions of eastern and central United States and in southeastern Canada. It is composed of small aquatic or semiaquatic turtles that are bottom crawlers. They are mainly carnivores and scavengers. Two modern genera, *Kinosternon* and *Sternotherus,* occur in North America today and both are represented by Pleistocene fossils. No extinct Pleistocene genera have been described.

Osteological characters include a fully developed shell with no plastral reduction. A cervical (nuchal) scute impression (see Fig. 22) occurs in the nuchal bone (Fig. 23A). The plastron is either doubly hinged (*Kinosternon*) or unhinged (*Sternotherus*). An entoplastral (see Fig. 22) bone is lacking. The costal bones have short, unstriated ribs. The peripheral bones are short. The limb bones lack striations.

Kinosternon Spix, 1824: Mud Turtles

These turtles are small animals with brownish or yellowish brown shells and large heads. One species has yellow stripes on the carapace and on the head. This genus has both aquatic and semiaquatic species, but all species prefer shallow, still water. They range throughout much of temperate North America south of Canada from the East Coast to the Rocky Mountains. All three of the

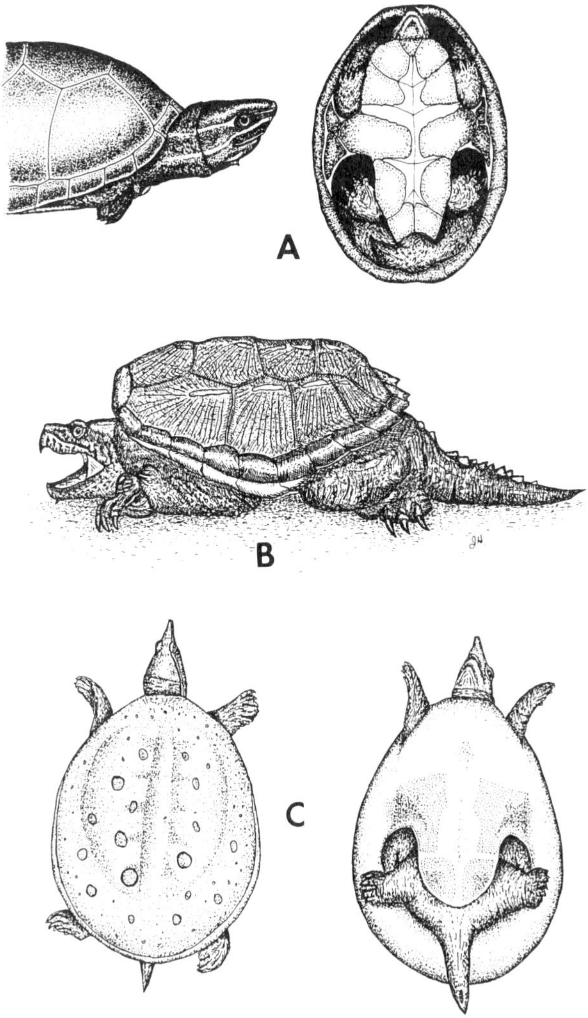

FIGURE 6. Representatives of North American turtle families. (A) *Sternotherus odoratus* (Kinosternidae). (B) *Chelydra serpentina* (Chelydridae). (C) *Apalone spinifera* (Trionychidae). These species occurred in the Pleistocene of North America. [Drawings by J. H. Harding in Holman and Harding, 1977]

modern North American species are known as fossils and no extinct Pleistocene species have been described.

Regional Pleistocene Occurrences

Kinosternon subrubrum. Common mud turtle: Appalachian and Southeastern regions.

Kinosternon baurii. Striped mud turtle: Southeastern region.

Kinosternon flavescens. Yellow mud turtle: Northern Plains and Southern Plains regions.

Sternotherus Gray, 1825: Musk Turtles

Musk turtles (Fig. 6A) are small animals with dull brownish or brownish green colored shells and often have yellow stripes on the head. They have large or very large heads. They are all aquatic and prefer still, shallow water. Modern species of this genus occur widely in the temperate part of about the eastern one-third of the United States and in southeastern Canada. Two of the four North American species are known as Pleistocene fossils, but no extinct Pleistocene species have been described.

Regional Pleistocene Occurrences

Sternotherus odoratus. Common musk turtle: Northeastern/Great Lakes, Southeastern, Northern Plains, and Southern Plains regions.

Sternotherus minor. Loggerhead musk turtle: Southeastern region.

Family Chelydridae Gray, 1870

Snapping Turtles (Figs. 6B, 23B)

Snapping turtles are large or very large animals with dull-colored, reduced shells, large or huge heads, and long tails. All of them walk along the bottom of lakes, ponds, and streams, and all are of carnivorous or scavenging habits. The family occurs in most of the United States today west to the Rocky Mountains and in southeastern and south-central Canada. Two genera, *Chelydra* and *Macroclemys,* occur in North America today and both are represented by Pleistocene fossils. No extinct Pleistocene genera have been described.

Osteological characters include a reduced shell with a small, cruciform plastron. The nuchal bone (Fig. 23B) lacks a cervical (nuchal) scute impression (see Fig. 22). The small plastron is unhinged and it lacks an entoplastral bone (see Fig. 22). Costal ribs are elongate and striated. Peripheral bones are relatively short and the limb bones are characteristically striated.

Chelydra Schweigger, 1812: Snapping Turtles

These large turtles have a blackish shell that lacks dorsal ridges and have a large head The genus consists of a single species, *Chelydra serpentina* (Fig. 6B), which occurs widely in the United States and in southern Canada east of the Rocky Mountains. It occurred west of the Rocky Mountains in Nevada in the Pleistocene. No extinct Pleistocene species are currently recognized.

Regional Pleistocene Occurrences

Chelydra serpentina. Snapping turtle: Northeastern/Great Lakes, Appalachian, Southeastern, Ozark, Northern Plains, Southern Plains, and Rocky Mountain/Great Basin regions.

Macroclemys Gray, 1855: Alligator Snapping Turtles

These huge turtles have a blackish shell with three high dorsal ridges and have a huge head. Alligator snappers have a tongue-lure that wiggles like a small worm and attracts fish to a gaping mouth. This genus is composed of a single species, *Macroclemys temmenckii*, which is today restricted to the southeastern

United States southward into the panhandle and northwestern portions of Florida. This species is also known as a Pleistocene fossil, but no extinct Pleistocene species are known.

Regional Pleistocene Occurrences

Macroclemys temmenckii. Alligator snapping turtle: Southeastern region.

Family Emydidae Lydekker, 1889

Emydid turtles (Figs. 7, 23D–H, 24A)

This large turtle family occurs throughout North America today except for the coldest regions. It is represented by nine genera that contain aquatic, semi-aquatic, and terrestrial species. Many of them are attractively colored with red or yellow patterns on the shell, neck, or head. All of the nine genera and seventeen of the twenty-seven modern North American species are represented as Pleistocene fossils. No extinct genera, but one extinct species, *Pseudemys hibbardi* (Fig. 36), and three extinct subspecies, *Terrapene carolina putnami* (Fig. 42A), *Trachemys scripta bisornata*, and *Trachemys scripta petrolei,* are recognized. These extinct forms will be addressed in the appropriate regional accounts.

Osteological characters include a fully developed shell with no plastral reduction. A cervical (nuchal) scute impression (see Fig. 22) occurs on the nuchal bone (Fig. 23D–H). The plastron is unhinged in most species, but some species have a single hinge that allows the shell to close around the entire animal. An entoplastral bone (see Fig. 22) is present. Costal ribs are short and unstriated. Peripheral bones are relatively long. Limb bones are unstriated.

Chrysemys Gray, 1844: Painted Turtles

Painted turtles (Fig. 7A) are medium-sized turtles with a blackish or olive-colored, very smooth shell, with reddish markings around the margin. The neck and head are attractively marked with yellow and red stripes. These animals prefer shallow, quiet water with ample aquatic vegetation. They are omnivorous in their feeding habits. Painted turtles are widely distributed in the United States except for about the southwestern one-third of the country, and they occur across southern Canada from coast to coast. The modern species, *Chrysemys picta*, is known as a Pleistocene fossil, but no extinct Pleistocene species are recognized.

Regional Pleistocene Occurrences

Chrysemys picta. Painted turtle: Northeastern/Great Lakes, Appalachian, Southeastern, Northern Plains, and Southern Plains regions.

Clemmys Ritgen, 1829: Marsh and Pond Turtles

This genus contains very small to large turtles with either rather plain blackish or brownish (Fig. 7C) or attractively spotted (Fig. 7B) shells. Most individuals have orange or reddish markings on the head. These animals are usually semiaquatic, but some of them spend quite a bit of time on land foraging for

FIGURE 7. Some North American turtles of the family Emydidae. (A) *Chrysemys picta* (B) *Clemmys guttata*. (C) *Clemmys insculpta*. (D) *Emydoidea blandingii*. (E) *Graptemys geographica*. (F) *Terrapene carolina*. These species occurred in the Pleistocene of North America. [Drawings by J. H. Harding in Holman and Harding, 1977]

food. They are omnivorous feeders. These animals occur mainly in the Great Lakes region of the United States and Canada, the eastern coastal states south into Florida, and again in the Pacific coastal states. The genus consists of four modern species, all of which are represented by Pleistocene fossils. No extinct Pleistocene species have been described.

Regional Pleistocene Occurrences

Clemmys guttata. Spotted turtle: Southeastern region.
Clemmys insculpta. Wood turtle: Appalachian region.
Clemmys muhlenbergii. Bog turtle: Appalachian region.
Clemmys marmorata. Western pond turtle: Pacific Coast region.

Deirochelys Agassiz, 1857: Chicken Turtles

These medium-sized turtles have a blackish reticulated shell and a very long yellow and greenish striped neck. They live in shallow, quiet aquatic situations where they obtain their food by a very effective gape-and-suck method of feeding. Chicken turtles occur mainly in the southeastern United States west into Texas. This genus consists of a single modern species, *Deirochelys reticularia*, which has been recorded as a Pleistocene fossil, but no extinct Pleistocene species are recognized.

Regional Pleistocene Occurrences

Deirochelys reticularia. Chicken turtle: Southeastern region.

Emydoidea Gray, 1870: Blanding's Turtles

Blanding's turtles (Fig. 7D) are large turtles with a black shell covered with many light spots, a very long neck, and a head with a bright yellow throat patch. They have a single hinge in the plastron, but different individuals and populations differ in their ability to close the shell. Blanding's turtles prefer still, shallow water and have an effective gape and suck method of feeding. This genus mainly occurs in the northeastern United States east to central Nebraska and in southeastern Canada today, but the Pleistocene range of the single modern species, *Emydoidea blandingii*, was much more extensive than it is at present. No extinct Pleistocene species have been described.

Regional Pleistocene Occurrences

Emydoidea blandingii. Blanding's turtle: Northeastern/Great Lakes, Appalachian, Southeastern, Ozark, Northern Plains, and Southern Plains regions.

Graptemys Agassiz, 1857: Map Turtles

Map turtles (Fig. 7E) are large turtles that have a maplike pattern on their greenish shells and yellow and greenish stripes on the neck and head. They are very aquatic species that prefer larger streams and lakes. These animals are mainly carnivorous and some are specialized for feeding upon mollusks. Map turtles are widely distributed in the eastern and central United States and occur in southeastern Canada. There are eleven modern species, three of which are known as Pleistocene fossils. No extinct Pleistocene species have been described.

Regional Pleistocene Occurrences

Graptemys geographica. Common map turtle: Appalachian and Northern Plains regions.

Graptemys barbouri. Barbour's map turtle: Southeastern region.

Graptemys pseudogeographica. False map turtle: Northern Plains and Southern Plains regions.

Additionally, fossils identified as *Graptemys* sp. have been reported from the Northeastern/Great Lakes region.

Malaclemys Gray, 1844: Diamondback Terrapins

These large turtles have diamond-shaped markings on the carapace and a light greenish, spotted head. They are mainly carnivorous and occur in brackish coastal waters of the United States from Cape Cod around the Gulf of Mexico to southern Texas. The single modern species, *Malaclemys terrapin*, has been reported as a Pleistocene fossil, but no extinct Pleistocene species have been described.

Regional Pleistocene Occurrences

Malaclemys terrapin. Diamondback terrapin: Southeastern region.

Pseudemys Gray, 1855: Cooters and Red-bellied Turtles

These large turtles have mainly dark-colored, rather attractively marked shells, and most have yellow stripes on the head and neck. They are aquatic turtles that prefer rather shallow, still water. They commonly are carnivores when they are young, but they may revert to a strictly herbivorous diet as they reach maturity. Their main range is in the southeastern and south-central United States. Five modern species are recognized, three of which have been identified as Pleistocene fossils. An extinct Pleistocene species, *Pseudemys hibbardi* (Fig. 36), has been described and will be discussed in the appropriate regional account.

Regional Pleistocene Occurrences

Pseudemys concinna. River Cooter: Appalachian, Southeastern, Northern Plains, and Southern Plains regions.

Pseudemys nelsoni. Florida red-bellied turtle: Appalachian and Southeastern regions.

Pseudemys floridana. Common cooter: Southeastern and Northern Plains regions.

Pseudemys hibbardi. Hibbard's cooter: Northern Plains and Southern Plains regions.

Additionally, fossils identified as *Pseudemys* sp. have been reported from the Northeastern/Great Lakes region.

Terrapene Merrem, 1820: Box Turtles

Box turtles (Fig. 7F) are medium-sized turtles that almost always have yellow or orange markings on their dome-shaped carapaces, and that can completely enclose the head and all the body parts within the shell. These omnivorous, terrestrial reptiles occur today in the temperate parts of the United States, mainly from the East Coast to the Rocky Mountains. Both of the two modern species have been recorded as Pleistocene fossils and no extinct Pleistocene species are currently recognized. Nevertheless, an extremely large extinct subspecies, *Terrapene carolina putnami* (Fig. 42A) has been recognized, as have supposed hybrids between the extinct subspecies and modern subspecies. This situation will be discussed later in this book.

Regional Pleistocene Occurrences

Terrapene carolina. Common box turtle: Northeastern/Great Lakes, Appalachian, Southeastern, Ozark, Northern Plains, Southern Plains, and Rocky Mountain/Great Basin regions.

Terrapene ornata. Ornate box turtle: Ozark, Northern Plains, Southern Plains, and Rocky Mountain/Great Basin regions.

Trachemys Agassiz, 1857: Sliders

These large turtles have a rather dull-colored shell, but have attractive yellow stripes on the neck and head, and sometimes a red blotch behind the eye. They are aquatic animals that prefer low-energy aquatic situations with abundant vegetation and are omnivorous. This genus is widely distributed in the southeastern and central United States and has been widely introduced to other parts of the world as the result of pet trade activities. The modern species, *Trachemys scripta*, is known as a Pleistocene fossil. Two extinct Pleistocene subspecies, *Trachemys scripta bisornata* and *Trachemys scripta petrolei*, have been described and will be discussed under the appropriate regional account.

Regional Pleistocene Occurrences

Trachemys scripta. Common slider: Appalachian, Southeastern, Ozark, Northern Plains, and Southern Plains regions.

Family Testudinidae Gray, 1825

Tortoises (Figs. 8, 43)

This family consists of land tortoises, most of which are strictly herbivorous. It is represented in North America today by a single genus, *Gopherus* (Fig. 8), which occurs in the southeastern and southwestern United States. But in the Pleistocene, several extinct species of another testudinid genus, *Geochelone* (Fig. 43), were widespread in the United States, as was at least one large extinct species of *Gopherus*.

Osteologically, the Testudinidae are very similar to the Emydidae. The major differences include the fact that the hind feet in the Testudinidae have a stumpy skeletal structure and sometimes the front feet are skeletally modified for burrowing, and that some species lack a cervical (nuchal) scute impression on

FIGURE 8. A representative of the North American turtle family Testudinidae. *Gopherus polyphemus* courtship behavior postures. This species occurred in the Pleistocene of North America. [From Auffenberg, 1966; permission of *Herpetologica*]

the nuchal bone. The Testudinidae have the same arrangement of the plastral bones as in the Emydidae (see Fig. 22), and neither family has striations on the ribs or limb skeleton.

Geochelone Fitzinger, 1835: Geochelone Tortoises

All North American species of *Geochelone* (Fig. 43) became extinct by the end of the Pleistocene, about 10,000 years ago. These extinct taxa consisted of both giant and smaller species. Often both large and small *Geochelone* have not been specifically identified in the scientific literature because of unpublished and continuing studies on the species of the genus and a somewhat unsettled taxonomy. I have taken a conservative approach in this book following Preston (1979) and recognize that more species potentially exist than are listed here.

Regional Pleistocene Occurrences

Geochelone crassiscutata. Eastern large tortoise: Northeastern/Great Lakes, Appalachian, and Southeastern regions.

Geochelone incisa. Eastern small tortoise: Southeastern region.

Geochelone mlynarskii. Mlynarski's small tortoise: Southeastern region

Geochelone "species A." Undetermined western large tortoise: Northern Plains and Southern Plains regions.

Geochelone "species B." Undetermined western small tortoise: Northern Plains and Southern Plains regions.

Geochelone equicomes. Western medium to large tortoise: Northern Plains region.

Geochelone wilsoni. Western small tortoise: Southern Plains and Rocky Mountain/Great Basin regions.

Additionally, turtle remains identified as "cf. *Geochelone*" have been reported from the Ozark region, and remains identified as "(?) *Geochelone*" have been reported from the Pacific Coast region.

Gopherus Rafinesque, 1832: Gopher Tortoises

These are large turtles (by modern standards) with dull-colored shells and stumpy hind feet. All the modern species make extensive burrows and are herbivorous. They mainly live in dry or desert areas in the southeastern and southwestern United States today. The genus is comprised of three modern and at least one large extinct species, *Geochelone laticuda*. Two of the three modern species are known as Pleistocene fossils. As in the case of *Geochelone* tortoises, fossil records of *Gopherus* are often reported as "*Gopherus* sp. indet." or "*Gopherus* large extinct species."

Regional Pleistocene Occurrences

Gopherus agassizii. Desert tortoise: Rocky Mountain/Great Basin and Pacific Coast regions.

Gopherus polyphemus. Gopher tortoise: Southeastern and Southern Plains regions.

Gopherus laticauda. Extinct burrowing tortoise: Southern Plains region.

Additionally, Pleistocene remains have been assigned to *Gopherus* sp. (this assumes one of the modern taxa) from the Southeastern, Northern Plains, Rocky Mountain/Great Basin, and Pacific Coast regions. *Gopherus*—"Undetermined large extinct species" has been reported from the Northern Plains and Southern Plains regions.

Family Trionychidae Bell, 1828

Softshell Turtles (Figs. 6C, 24B)

Softshell turtles in modern North America consist of a single genus, *Apalone* (Fig. 6C), which contains large turtles with a flat, greenish or brownish, soft, leathery shell, long necks, a long snorkellike nose, and highly webbed feet. They are very aquatic animals that live in permanent rivers, streams, lakes, and sometimes ponds. They are carnivores that often prefer crayfish. They are widely distributed in temperate areas in the United States today except for the Rocky Mountain and Pacific areas, and they also occur in southeastern Canada. No extinct Pleistocene genera have been described.

Osteological characters include a much reduced bony shell, both above and below, that is covered with a pliable leathery skin. A cervical (nuchal) scute impression (see Fig. 22) is lacking in the simple nuchal bone. The reduced plastron lacks an entoplastral bone (see Fig. 22). Costal ribs are elongate and characteristically striated. Peripheral bones are lacking. Limb bones are also characteristically striated. The individual carapacial bones are ornamented with a pit and ridge design dorsally (Fig. 24B).

Apalone Rafinesque, 1832: Softshell Turtles

Three modern species occur natively in North America today, and two of them have been reported as Pleistocene fossils. No extinct Pleistocene species have been described.

Regional Pleistocene Occurrences

Apalone spinifera. Spiny softshell turtle: Northeastern/Great Lakes, Appalachian, Northern Plains, and Southern Plains regions.
Apalone ferox. Florida softshell turtle: Southeastern region.

Order Sauria McCartney, 1802: Lizards

Ten families of lizards occur natively in the United States today and nine of them are represented by Pleistocene fossils. Only the Family Xantusiidae is absent from the North American Pleistocene fossil record. These nine families are:

Crotaphytidae (collard lizards)
Iguanidae (iguanids)
Phrynosomatidae (phrynosomatids)
Polychridae (polychridids)
Gekkonidae (geckos)
Teiidae (whiptails and racerunners)
Scincidae (skinks)
Anguidae (alligator and glass lizards)
Helodermatidae (venomous lizards)

Of the twenty genera and forty-five species of lizards reported from the Pleistocene of North America, only one species, *Eumeces carri*, from the early Irvingtonian, is recognized as extinct. This extinct species will be addressed under the appropriate regional account.

The following four families (and four others not listed here) were, until recently, considered a single family, the Iguanidae. All of these families have many generalized osteological characters, including procoelous vertebrae, a parietal foramen in the skull, and a lack of osteoscutes.

Family Crotaphytidae Smith and Brodie, 1982

Collard and Leopard Lizards: (Fig. 9A, 25G)

This family is represented in North America today by two genera, both of which occur mainly in the southwestern and far western United States. These are large, alert, brightly marked lizards that feed upon large invertebrates and occasionally upon smaller lizards and snakes. Both of these genera are known as Pleistocene fossils and no extinct Pleistocene genera have been described.

Osteological characters include a poorly developed or absent labial blade of

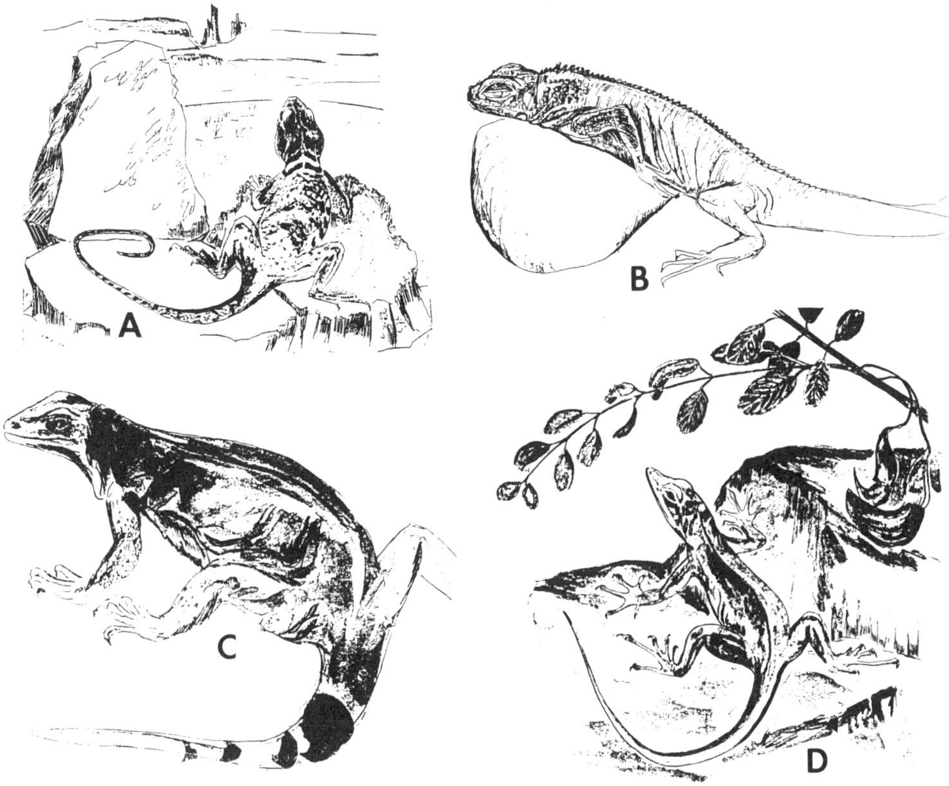

FIGURE 9. Representatives of North American iguanine Lizards. (A) *Crotaphytus collaris* (Crotaphytidae). (B) *Ctenosaura pectinata* (Iguanidae). (C) *Sauromalus obesus* (Iguanidae). (D) *Anolis carolinensis* (Polychridae). Species A, C, and D occurred in the Pleistocene of North America. [From watercolors by D. R. Holman]

the coronoid. The Meckel's groove in the dentary is not fused (Fig. 25G). The splenial is relatively long anteriorly. Palatine teeth are present and the posterior process of the interclavicle is not invested by the sternum far anteriorly. Four sternal ribs are present.

Crotaphytus Holbrook, 1842: Collard Lizards

These are large lizards (Fig. 9A) with a large, yellowish head, a black neck collar, and greenish limbs. This genus occurs mainly in the southwestern United States. Both modern species of this genus are known as Pleistocene fossils, but no extinct Pleistocene species have been recognized.

Regional Pleistocene Occurrences

Crotaphytus collaris. Collard lizard: Southern Plains and Rocky Mountain/Great Basin regions.

Crotaphytus reticulatus. Reticulate collard lizard: Southern Plains and Rocky Mountain/Great Basin regions.

Additionally, *Crotaphytus* sp. has been recognized in the Ozark and the Pacific Coast regions.

Gambelia Baird, 1859: Leopard Lizards

These are large, spotted lizards with large heads. This genus occurs mainly in the far western United States from western Texas to the Pacific states. There are two leopard lizard species, one of which, *Gambelia wislizenii*, is known as a Pleistocene fossil. No extinct Pleistocene species have been described.

Regional Pleistocene Occurrences

Gambelia wislizenii. Long-nosed leopard lizard: Rocky Mountain/Great Basin and Pacific Coast regions.

Family Iguanidae Oppel, 1811

Iguanid Lizards (Fig. 9C)

This family is represented in the United States and Canada today by three genera. *Dipsosaurus* and *Sauromalus* (Fig. 9C) live in the far western United States, and *Ctenosaura* (Fig. 9B) just gets into southern Texas. These are large lizards of herbivorous or omnivorous habits. Only *Sauromalus* has been recorded as a North American Pleistocene fossil. No extinct Pleistocene iguanid genera have been described.

Osteological characters of the family include a large labial blade of the coronoid. The Meckel's groove in the dentary is fused. The splenial is relatively short anteriorly. Palatine teeth are absent. The posterior process of the interclavicle is not invested by the sternum anteriorly and there are four sternal ribs.

Sauromalus Duméril, 1856: Chuckwallas

Chuckwallas are large, dark-colored lizards with loose folds of skin on the sides of the body and neck. They are herbivores that live in very dry rocky regions, where they often inflate their bodies and wedge themselves between rocks for protection against predators. The single species that occurs in the area covered by this book, *Sauromalus obesus* (Fig. 9C), is also known as a Pleistocene fossil. No extinct Pleistocene species have been described.

Regional Pleistocene Occurrences

Sauromalus obesus. Chuckwalla: Rocky Mountain/Great Basin and Pacific Coast regions.

Family Phrynosomatidae Blainville, 1835

Phrynosomatid Lizards (Fig. 25C,F)

This large family occurs from coast to coast in North America and in extreme southwestern Canada. They are mainly insectivorous forms that occur in a variety of sizes and shapes. Ten genera are known in North America in the area covered by this book and seven of them are represented by Pleistocene fossils. No extinct Pleistocene genera have been described.

Osteological characters include a poorly developed or absent labial blade of the coronoid. The Meckel's groove is open (Fig. 25C). The splenial is relatively long anteriorly and palatal teeth are absent. The posterior process of the interclavicle is invested by the sternum far anteriorly and there are either three or four sternal ribs.

Callisaurus Blainville, 1835: Zebra-Tailed Lizards

These medium-sized lizards have dark stripes on the side of the chest and are very long-legged and slim. They are capable of running very fast. The genus is represented by a single living species, *Callisaurus draconoides*, which occurs in the far southwestern United States. This species is represented by Pleistocene fossils, but no extinct Pleistocene species have been described.

Regional Pleistocene Occurrences

Callisaurus draconoides. Zebra-tailed lizard: Rocky Mountain/Great Basin region.

Cophosaurus Troschel, 1852: Greater Earless Lizards

These medium-sized lizards have dark stripes on the side of the groin and are slimly built forms with thin legs and compressed tails. This genus occurs in dry habitats of the southwestern United States today and is represented by a single modern species, *Cophosaurus texanus,* which also occurs as a Pleistocene fossil. No extinct Pleistocene species have been described.

Regional Pleistocene Occurrences

Cophosaurus texanus. Greater earless lizard: Southern Plains and Rocky Mountain/Great Basin regions.

Holbrookia Girard, 1851: Earless Lizards

These are small terrestrial lizards with a short tail that is unmarked below. These animals live in the plains and southwestern parts of the United States today. Only one of the three modern species, *Holbrookia maculata*, is represented by Pleistocene fossils. No extinct Pleistocene species have been described.

Regional Pleistocene Occurrences

Holbrookia maculata Lesser earless lizard. Rocky Mountain/Great Basin region.

Additionally, *Holbrookia* sp. has been reported from the Northern Plains and Southern Plains regions.

Phrynosoma Weigmann, 1828: Horned Lizards

This genus is composed of mainly brownish or grayish, flattened horned lizards with short limbs and a very short tail. Ants are the preferred food of many of the species. The genus occurs from the Great Plains westward to the Pacific Coast in the United States and in extreme southwestern Canada. There are seven modern North American species, six of which are represented by Pleistocene fossils. No extinct Pleistocene species have been described.

Regional Pleistocene Occurrences

Phrynosoma douglassii. Short-horned lizard: Northern Plains and Rocky Mountain/Great Basin regions.

Phrynosoma cornutum. Texas horned lizard: Southern Plains and Rocky Mountain/Great Basin regions.

Phrynosoma modestum. Round-tailed horned lizard: Rocky Mountain/Great Basin region.

Phrynosoma platyrhinos. Desert horned lizard: Rocky Mountain/Great Basin region.

Phrnosoma solare. Regal horned lizard: Rocky Mountain/Great Basin region.

Phrynosoma coronatum. Coast horned lizard: Pacific Coast region.

Sceloporus Wiegmann, 1828: Spiny Lizards

Spiny lizards are medium-sized to large animals that are usually grayish to brownish in color. They are variously striped or have other dark markings on the body, and some species have black neck collars. The have keeled, pointed scales, and they occupy a variety of habitats. Spiny lizards occur from coast to coast in the temperate regions of the United States. In some southwestern areas two or more species occur together in the same locality. Sixteen modern species comprise this genus in North America today and nine of these are represented by Pleistocene fossils. No extinct Pleistocene species have been described.

Regional Pleistocene Occurrences

Sceloporus undulatus. Eastern fence lizard: Appalachian, Southeastern, Southern Plains, Rocky Mountain/Great Basin, and Pacific Coast region.

Sceloporus olivaceus. Texas spiny lizard: Southern Plains region.

Sceloporus poinsettii. Crevice spiny lizard: Southern Plains region.

Sceloporus variabilis. Rose-bellied lizard: Southern Plains region.

Sceloporus clarkii. Clark's spiny lizard: Rocky Mountain/Great Basin region.

Sceloporus graciosus. Sagebrush lizard: Rocky Mountain/Great Basin and Pacific Coast regions.

Sceloporus jarrovii. Yarrows spiny lizard: Rocky Mountain/Great Basin region.

Sceloporus magister. Desert spiny lizard: Rocky Mountain/ Great Basin and Pacific Coast regions.

Sceloporus occidentalis. Western fence lizard: Rocky Mountain/Great Basin and Pacific Coast regions.

Additionally, *Sceloporus* sp. has been identified as a fossil from the Northern Plains Region.

Urosaurus Hallowell, 1854: Tree Lizards

This genus contains small, plain-colored, arboreal lizards with a streaky pattern. They occur in the far southwestern United States today. Two of the modern species have been identified as Pleistocene fossils, but no extinct Pleistocene species have been described.

Regional Pleistocene Occurrences

Urosaurus graciosus. Long-tailed brush lizard: Rocky Mountain/Great Basin region.

Urosaurus ornatus. Tree lizard: Rocky Mountain/Great Basin region.

Uta Baird and Girard, 1852: Side-Blotched Lizards

Side-blotched lizards are small, plain-colored animals that have a dark bluish blotch on each side of the body behind the front limbs. They occur virtually ubiquitously from southwestern Texas northwestward to the Pacific Coast states. The single species in the genus, *Uta stansburiana*, has been identified as a Pleistocene fossil, but no extinct Pleistocene species of *Uta* have been described.

Regional Pleistocene Occurrences

Uta stansburiana. Side-blotched lizard: Rocky Mountain/Great Basin and Pacific Coast regions.

Family Polychridae Fitzinger, 1843

Polychridid Lizards (Figs. 9D, 25A)

This very specifically diverse family occurs mainly in tropical America today. These are mainly arboreal animals with brightly colored throat fans. The ability to change color from gray or brown to bright green is well-developed in most of these lizards. The family is mainly composed of a single genus, *Anolis*, which has more than 250 species. Only one acrobatic species, *Anolis carolinensis* (Fig. 9D), a small lizard that can rapidly change from gray or brown to dark green and that has a bright pink throat fan, occurs natively in the area covered by this book.

Osteological characters of the family include a labial blade of the coronoid that varies from well-developed to absent. Meckel's groove on the dentary is fused (Fig. 25A). The splenial is relatively short anterior to the level of the coronoid. Palatine teeth are usually present. The posterior process of the clavicle is variably invested by the sternum far anteriorly. There may be two, three, or four sternal ribs.

Anolis Daudin, 1802: Anoles

The single native North American species, *Anolis carolinensis* (Fig. 9D), occurs in the southern part of the Appalachian region and in the southeastern United States south to the Florida keys today. This species has been identified as a Pleistocene fossil, but no extinct Pleistocene species of the genus have been described.

Regional Pleistocene Occurrences

Anolis carolinensis. Green anole: Appalachian and Southeastern regions.

Family Gekkonidae Gray, 1825

Geckos (Fig. 10A)

This is a very large family with a worldwide, mainly tropicolitan distribution. These animals have soft skins and tiny scales, toe pads for climbing, large eyes with vertical pupils, and some of them are quite vocal. Many species do well around human dwellings in the tropics.

Osteological characters include large vomers that are broad and platelike. The parietal foramen is absent. Palatine and pterygoid teeth are absent, but there is a high mandibular tooth number. The vertebrae are amphicoelous and osteoderms are mainly lacking. Only two genera occur natively in the area

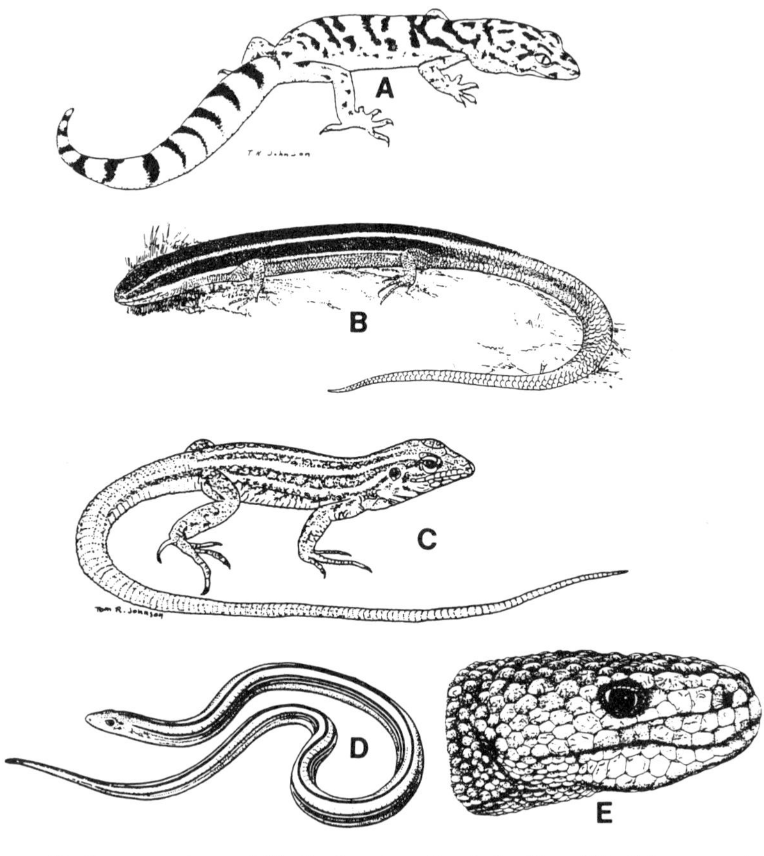

FIGURE 10. Representatives of North American lizard families. (A) *Coleonyx variegatus* (Gekkonidae). (B) *Cnemidophorus gularis* (Teiidae). (C) *Eumeces skiltonianus* (Scincidae). (D) *Ophisaurus attenuatus* (Anguidae). (E) *Heloderma horridum*. All these species and the genus *Heloderma* occurred in the Pleistocene of North America. [A and C, drawings by T. R. Johnson, permission of *Herpetological Review;* B from Van Denburgh, 1895; D by permission of Macmillan College Publishing Company from *Vertebrate Life* by F. H. Pough, J. B. Heiser, and W. N. McFarland. Copyright 1989 by Macmillan College Publishing Company, Inc.; E from Campbell and Vannini, 1988, permission of *Herpetological Journal*]

covered by this book and only one, *Coleonyx* (Fig. 10A), has been identified as a Pleistocene fossil. No extinct Pleistocene gecko genera have been described in North America.

Coleonyx Gray, 1845: Banded Geckos

These are small, delicate, banded lizards that occur in dry habitats in southwest Texas and the far southwestern United States. Four modern species are known in the United States, but only one of these, *Coleonyx variegatus* (Fig. 10A), has been identified as a Pleistocene fossil. No extinct Pleistocene species have been described.

Regional Pleistocene Occurrences

Coleonyx variegatus. Western banded gecko: Rocky Mountain/Great Basin region.

Family Teiidae Gray, 1827

Whiptails and Racerunners (Fig. 10C)

This very large family is especially diverse in South America, where some very large species are found. In North America, teiid lizards are represented by a single native genus, *Cnemidophorus* (Fig. 10C), which contains medium sized lizards that may be striped, spotted, or have a checkered pattern. They are active, nervous lizards that usually occur in dry, sandy habitats. They have tiny scales on their backs and large, rectangular scales on their bellies. Some species exist as populations of uniparental, all-female populations. *Cnemidophorus* ranges from coast to coast in temperate areas of the United Sates today in the area covered by this book. This genus occurs as a Pleistocene fossil and no extinct Pleistocene teiid genera have been described.

Osteological characters of the family include an elongate facial region of the skull. The vomer is elongate. The maxilla is excluded from the suborbital fenestra by the contact of the ectopterygoid and the palatine. The vertebrae have zygosphenal and zygantral articular surfaces. The dentition is usually heterodont, with a diastema occurring between the premaxilla and the maxilla. The teeth usually have an extensive cementum on their bases.

Cnemidophorus Wagler, 1830: Whiptails and Racerunners

Three of the sixteen or more species of *Cnemidophorus* are known as Pleistocene fossils. No extinct Pleistocene species of this genus have been described.

Regional Pleistocene Occurrences

Cnemidophorus sexlineatus. Six-lined raceruner: Southeastern, Ozark, Northern Plains, and Southern Plains regions.

Cnemidophorus gularis. Texas spotted whiptail: Southern Plains region.

Cnemidophorus tigris. Western whiptail: Rocky Mountain/Great Basin and Pacific Coast regions.

Family Scincidae Gray, 1825

Skinks (Figs. 10B, 25D,E)

Skinks are very widely distributed in temperate and tropical areas of the world. They are typically shiny lizards with short limbs that may have brightly colored heads or tails at certain times in their ontogeny. Some species have limbs that are greatly reduced or absent. The family is widespread from coast to coast in the United States today and also occurs in southeastern and southwestern Canada. Only three genera, *Eumeces* (Fig. 10B) *Neoseps,* and *Scincella,* occur natively in the area covered by this book, and *Eumeces* and *Scincella* are known as Pleistocene fossils. No extinct genera of Pleistocene skinks have been described.

Osteological characters of this family include a secondary palate that occurs in varying degrees of development. The palatine and pterygoid bones meet at the midline. The palate is usually toothless. Osteoscutes are present and they are cycloid in shape. The tooth crowns are often terminally striated (Fig. 25D) and they are concave lingually. The tooth crowns are terminally pigmented in at least one North American species.

Eumeces Wiegmann, 1834: Holarctic Skinks

These medium-sized to rather large lizards have shiny bodies, short legs, and relatively large heads. Males usually have more brightly colored heads than females and the tails of the subadults may be brightly colored. Thirteen modern species of skinks occur in North America today and eight of these are represented by Pleistocene fossils. One extinct Pleistocene species, *Eumeces carri*, has been described, and it will be discussed under the appropriate regional account.

Regional Pleistocene Occurrences

Eumeces fasciatus. Five-lined skink: Appalachian, Southeastern, Ozark, Northern Plains, and Southern Plains regions.

Eumeces laticeps. Broad-headed skink: Appalachian region.

Eumeces inexpectatus. Southeastern five-lined skink: Southeastern region.

Eumeces carri. Carr's skink: Southeastern region.

Eumeces obsoletus. Great Plains skink: Northern Plains, Southern Plains, and Rocky Mountain/Great Basin regions.

Eumeces septentrionalis. Prairie skink: Northern Plains region.

Eumeces tetragrammus. Four-lined skink: Southern Plains region.

Eumeces gilberti. Gilbert's skink: Pacific Coast region.

Eumeces skiltonianus. Western skink: Pacific Coast region.

Scincella Mittleman, 1950: Ground Skinks

Ground skinks are tiny brown lizards with dark stripes and long tails. They do not have bright heads and tails and are often found in the leaf litter. They mainly occur in the southeastern United States west to Texas and Oklahoma. Only one species, *Scincella lateralis,* occurs in North America today and it has been identified as a Pleistocene fossil. No extinct Pleistocene species have been described.

Regional Pleistocene Occurrences
Scincella lateralis. Ground skink: Appalachian region.

Family Anguidae Gray, 1825

Alligator and Glass Lizards (Fig. 10D, 25H)

This is a Holarctic family of elongate, shiny lizards whose limbs are reduced or absent. The family is represented in the area covered by this book by two genera, *Gerrhonotus* and *Ophisaurus* (Fig. 10D), both of which occur as Pleistocene fossils. No extinct Pleistocene anguid genera have been described.

Osteological characters of this family include a supratemporal arch that is reduced. The supratemporal fenestra is narrow. The mandibular bones are individually highly modified. The teeth are often recurved and chisel shaped, with the cutting edges forming an inwardly directed V. Osteoscutes are present, with the ventral ones being rectangular in shape. Vertebrae (Fig. 25H) are simple in structure.

Gerrhonotus Wiegmann, 1828: Alligator Lizards

These are large, generally light brownish, very elongate lizards with small limbs, a plated appearance, and shiny scales. They occur in the southwestern and far western United States and are represented by five modern species, only one of which, *Gerrhonotus multicarinatus*, is represented by Pleistocene fossils. No extinct Pleistocene species of this genus have been described.

Regional Pleistocene Occurrences
Gerrhonotus multicarinatus. Southern alligator lizard: Pacific Coast region.

Additionally, *Gerrhonotus* sp. has been identified from the Southeastern region in the early Irvingtonian in Florida, well out of its present range.

Ophisaurus Daudin, 1803: Glass Lizards

The glass lizards (Fig. 10D) are rather large, shiny, limbless lizards with long, fragile tails. They may be striped or spotted or both. Glass lizards mainly occur in the southeastern states and the temperate parts of the midwestern and plains states. Four modern species occur in the area covered by this book and three of them are represented by Pleistocene fossils. No extinct Pleistocene species have been described.

Regional Pleistocene Occurrences
Ophisuarus attenuatus. Slender glass lizard: Northeastern/Great Lakes, Ozark, and Northern Plains regions.
Ophisaurus ventralis. Eastern glass lizard: Southeastern and Ozark regions.
Ophisaurus compressus. Island glass lizard: Southeastern region.

In addition, *Ophisaurus* sp. was identified from the Southern Plains region.

Family Helodermatidae Wiegmann, 1829
Venomous Lizards (Fig. 10E)

These animals are large, heavy-bodied lizards with beaded scales and a massive head (Fig. 10E). They inject venom by chewing it in with the teeth. This family is represented in the United States and Canada by a single species, *Heloderma suspectum*, a very large lizard with beaded scales that occur in a pattern of orange, yellow, and black. This species occurs mainly in succulent desert habitats where it feeds on rodents, small birds, and eggs of other amniotes. This lizard occurs only in southern New Mexico, Arizona, Nevada, and California.

Osteological characters of the family include a short and broad skull. The temporal arch is absent. The premaxillary has a very large incisive process. The parietal foramen is absent. The teeth all have at least incipient venom channels. Osteoscutes are polygonal and subconical.

Heloderma Wiegmann, 1829: Gila Monsters and Beaded Lizards
The only species that occurs in the area covered by this book, *Heloderma suspectum*, is also known as a Pleistocene fossil. No extinct Pleistocene species have been described.

Regional Pleistocene Occurrences
Heloderma suspectum. Gila monster: Rocky Mountain/Great Basin region.

Order Amphisbaenia Gray, 1844: Worm Lizards

These mainly tropicolitan animals are either legless or have an obsolete pair of front limbs only. They have annular rings and tiny eyes and generally resemble earthworms.

Family Rhinuridae Vanzolini, 1951
Worm Lizards

A single monotypic family, the Rhinuridae, occurs in the area covered by this book. This family is represented by a single species, *Rhineura floridana*, a tiny, pinkish, wormlike animal, with large head scales that are obvious only at close range. It is a rarely seen fossorial form that prefers loose soils and occurs only in peninsular Florida today. No extinct Pleistocene worm lizard genera have been described.

The osteological characters for the family include a strong angle between the cranial and facial region of the skull. The parietals are fused, with an anterior process that separates the posterior ends of the frontals. The surangular bone is bent sharply downward. The premaxillary tooth count is reduced to one to three teeth. The vertebrae are procoelous and simple with the bottom of the centrum unkeeled and smooth. Numerous longitudinal ridges occur dorsally on the neural arches of the vertebrae.

Rhineura Cope, 1861: North American Worm Lizards

The single modern species, *Rhineura floridana*, also occurs as a Pleistocene fossil. No extinct Pleistocene species of this genus have been described.

Regional Pleistocene Occurrences

Rhineura floridana. Florida worm lizard: Southeastern region.

Order Serpentes Linnaeus, 1758: Snakes

Five families of snakes occur natively in mainland United States and Canada today and all of them are represented by Pleistocene fossils in the area. These five families are:

Boidae (boas and pythons)
Leptotyphlopidae (slender blind and thread snakes)
Colubridae (colubrid snakes)
Elapidae (elapids)
Viperidae (pit vipers and vipers)

Thirty-seven genera and seventy-eight species of snakes have been reported as North American Pleistocene fossils. None of these genera are extinct, but four extinct species, *Diadophis elinorae*, *Lampropeltis intermedius*, *Regina intermedia*, and *Crotalus potterensis,* have been described. These extinct taxa will be discussed in the appropriate regional accounts.

Family Boidae Gray, 1825

Boas and Pythons (Figs. 11A, 26C)

Boas and pythons are mainly very large, heavy-bodied snakes, and all of them kill their prey by constriction. They are mainly of tropicolitan occurrence. The North American boid genera, however, are small to medium-sized forms. Only two genera are known, *Charina* and *Lichanura*, and only the latter genus is known as a North American Pleistocene fossil.

Osteological characters of the family include a quadrate bone that is short and in a vertical position in the skull. The postfrontal is absent. An optic foramen occurs between the frontal and parietal bones. The vomerine process of the premaxilla is reduced and is in loose contact or separated from the vomers. Vestiges of the pelvic girdle are almost always present in the males. The vertebrae are relatively simple in structure. They are short and wide with massive zygosphenal and zygantral articulations and sometimes with a short, massive neural spine (Fig. 26C). Hypapophyses (see Fig. 26A) are present only on the anterior trunk vertebrae in most groups.

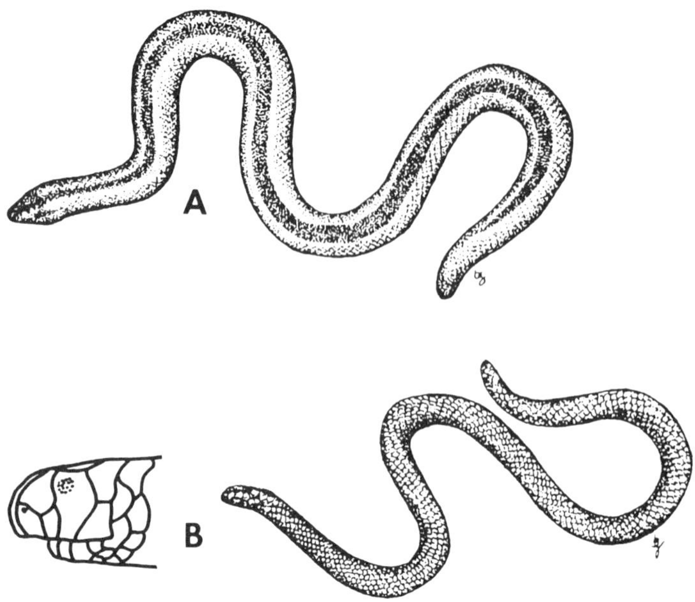

FIGURE 11. Representatives of primitive North American snake families. (A) *Lichanura trivirgata* (Boidae). (B) *Leptotyphlops dulcis* (Leptotyphlopidae). Both these genera and the species *Lichanura trivirgata* occurred in the Pleistocene of North America. [A and B (body), drawings by B. Gudgeon in Holman, 1976; B (head) from Dowling, 1959; permission of *Copeia*]

Lichanura Cope, 1861: Rosy Boas

Rosy boas (Fig. 11A) are medium-sized, constricting snakes with reddish brown stripes that inhabit rocky shrub and desert lands. They feed mainly on lizards and rodents. In the area covered by this book, rosy boas occur only in southwestern Nevada and southern California. The genus is represented in North America by a single species, *Lichanura trivirgata*. This species is represented by Pleistocene fossils, but no extinct Pleistocene species have been described.

Regional Pleistocene Occurrences

Lichanura trivirgata. Rosy boa: Pacific Coast region.

Family Leptotyphlopidae Fitzinger, 1843

Slender Blind and Thread Snakes (Fig. 11B)

These tiny snakes with degenerate eyes are widely spread in the neotropics, but in the area covered by this book they are represented by only a single genus, *Leptotyphlops* (Fig. 11B), a tiny purplish snake with a blunt tail that resembles the head, and eyes that look like two dots. These are secretive snakes that feed on ants and termites. This genus is known as a Pleistocene fossil and no extinct Pleistocene leptotyphlopid genera have been described.

Osteological characters of the family include a very solidly built skull that lacks the kineticism of most snake groups. The vertebrae are unique in that they lack neural spines, the neural arches are highly depressed, hypapophyses (see Fig. 26A) are limited to only two to four anterior vertebrae, and the centrum is smooth and unkeeled.

Leptotyphlops Fitzinger, 1843: Slender Blind Snake

This genus is represented in the United States by two species which live in the southwest. Only one of these, *Leptotyphlops humilis*, is represented by Pleistocene fossils. No extinct Pleistocene species of this genus have been described.

Regional Pleistocene Occurrences

Leptotyphlops humilis. Western slender blind snake: Rocky Mountain/Great Basin region.

Family Colubridae Oppel, 1811

Colubrids (Figs. 12, 26, 27, 28A,B, 31, 38)

This huge family of advanced snakes is very ecologically diverse. It includes degenerate burrowing forms, water snakes, vine and tree snakes, large constrictors, tiny worm and slug specialists, and some rear-fanged, venomous species. The group is considered to be polyphyletic. This family is represented

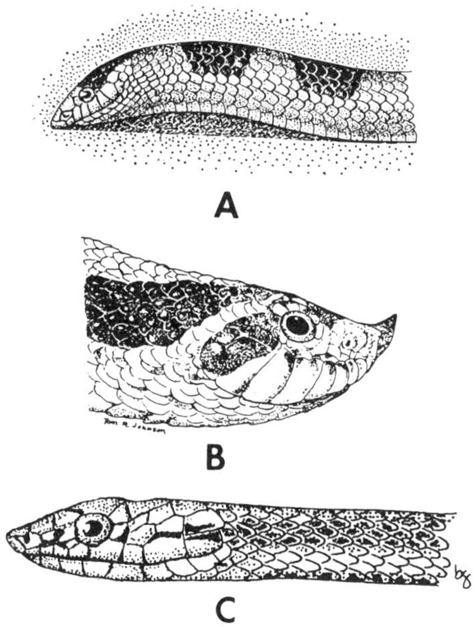

FIGURE 12. Representative North American snakes of the family Colubridae. (A) *Chionactis occipitalis*. (B) *Heterodon simus*. (C) *Masticophis taeniatus*. All these species occurred in the Pleistocene of North America. [A from Norris and Kavanau, 1966; permission of *Copeia*; B, drawing by T. R. Johnson, permission of *Herpetological Review*; C, drawing by B. Gudgeon in Holman, 1976]

in the Pleistocene by thirty-one genera and sixty-two species in the area covered by this book. No extinct genera have been described, but three extinct species, *Diadophis elinorae* (Fig. 31), *Lampropeltis intermedius* (Fig. 38), and *Regina intermedia* are currently recognized. These will be discussed in the appropriate regional accounts.

Osteological characters of the family include the fact that hollow poison fangs are not present and the maxilla is not retractable. No vestiges of limb girdles are present. The quadrate bone is elongate and it is not vertically oriented. Teeth are present on several bones in the cranium as well as on the dentary and maxillary bones. The vertebrae are usually longer than wide and lightly built with nonmassive zygosphenal and zygantral articular surfaces. Neural spines are long and usually relatively thin. Hypapophyses occur only in the anterior vertebrae in most groups, but in some taxa they occur on all the trunk vertebrae.

Arizona Kennicott, 1859: Glossy Snakes

These are medium sized, lightly spotted snakes with a glossy appearance. They occur in quite a variety of open habitats, where they feed on lizards, snakes, and small mammals that they kill by constriction. Glossy snakes occur mainly in the southwestern states today. The genus consists of a single modern species, *Arizona elegans*, which is known as a Pleistocene fossil. No extinct Pleistocene species of this snake have been described.

Regional Pleistocene Occurrences

Arizona elegans. Glossy snake: Southern Plains, Rocky Mountain/Great Basin, and Pacific Coast regions.

Bogartophis Brown, 1901: Trans-Pecos Rat Snake

These are large, yellowish brown snakes with a streaky black pattern. They are mainly nocturnal forms that feed on lizards, birds, and small mammals that they kill by constriction. In the area of this book, these animals occur only in southwestern Texas and southern New Mexico today. This genus consists of a single species, *Bogartophis subocularis*, which is also known as a Pleistocene fossil. No extinct Pleistocene species of this genus have been described.

Regional Pleistocene Occurrences

Bogartophis subocularis. Trans-Pecos rat snake: Southern Plains and Rocky Mountain/Great Basin regions.

Carphophis Gervais, 1843: Worm Snakes

These are very small, shiny, brownish snakes that generally resemble earthworms. They are secretive, woodland forms that mainly feed on earthworms and other soft-bodied invertebrates Worm snakes occur in the temperate parts of the eastern and central United States. The genus consists of a single modern species, *Carphophis amoenus*, that is also known as a Pleistocene fossil. No extinct Pleistocene species have been described.

Regional Pleistocene Occurrences

Carphophis amoenus. Worm snake: Appalachian, Southeastern, and Ozark regions.

Cemophora Cope, 1860: Scarlet Snakes

These small snakes are red, yellow, and black banded mimics of the coral snake. These colorful snakes are constrictors that occur in loose soils and feed on young mice, small snakes, small lizards, and eggs of other small amniotes. Scarlet snakes occur mainly in the southeastern states, reaching Texas and southern Indiana. The genus consists of a single modern species, *Cemophora coccinea*. The living species has been identified as a Pleistocene fossil, but no extinct Pleistocene species have been described.

Regional Pleistocene Occurrences

Cemophora coccinea. Scarlet snake: Appalachian and Southeastern regions.

Chionactis Cope, 1860: Shovel-Nosed Snakes

These colorful little snakes have red and black blotches on a yellowish ground color and a scooplike rostrum (Fig. 12A). They are burrowing, nocturnal forms that feed on invertebrates. Shovel-nosed snakes occur in southern Arizona, Nevada, and California, and are represented by three species, one of which, *Chionactis occipitalis* (Fig. 12A), is known as a Pleistocene fossil. No extinct Pleistocene species have been described.

Regional Pleistocene Occurrences

Chionactis occipitalis. Western shovel-nosed snake: Rocky Mountain/Great Basin and Pacific Coast regions.

Coluber Linnaeus, 1758: Racers

These are large, slim, black or bluish black, or yellowish snakes that are ubiquitous in their choice of habitats. Racers are alert, fast, nonconstricting snakes that feed on frogs, lizards, small snakes, and small mammals. Racers occur from coast to coast in many temperate parts of the United States and also occur in southeastern and southwestern Canada. Only a single, widespread species, *Coluber constrictor*, is currently recognized in the modern fauna. This species is also widespread as a Pleistocene fossil. No extinct Pleistocene species of *Coluber* have been described.

Regional Pleistocene Occurrences

Coluber constrictor. Racer: Northeastern/Great Lakes, Appalachian, Southeastern, Ozark, Northern Plains, Southern Plains, Rocky Mountain/Great Basin, and Pacific Coast regions.

Diadophis Baird and Girard, 1853: Ring-Necked Snakes

These are small blackish snakes with a light ring around their necks and a bright reddish or yellowish belly. They are secretive woodland forms that eat small salamanders, worms, other soft-bodied invertebrates, and sometimes tiny snakes.

They occur in southeastern Canada and most of the United States except for parts of the Rocky Mountain and Great Basin areas. This genus is represented by a single modern species, *Diadophis punctatus*, which is also known widely as a Pleistocene fossil. An extinct Pleistocene species of ring-necked snake, *Diadophis elenorae* (Fig. 31), has been described and will be discussed in the appropriate regional section.

Regional Pleistocene Occurrences

Diadophis punctatus. Ring-necked snake: Appalachian, Southeastern, Ozark, Northern Plains, Southern Plains, Rocky Mountain/Great Basin, and Pacific Coast regions.

Diadophis elenorae. Florida ring-necked snake: Southeastern region.

Drymarchon Fitzinger, 1843: Indigo Snakes

These are very large, shiny, bluish black snakes that often seek shelter in the burrows of gopher tortoises. They often occur in dry habitats, but in some areas they occur in dense, semitropical hammocks. Indigo snakes are nonconstricting species that feed mainly on frogs, snakes (sometimes poisonous ones), and small mammals. In the area of this book, these snakes occur only in Florida and southern parts of Georgia, Alabama, and Mississippi. Only a single species, *Drymarchon corais*, occurs in the modern fauna. The modern species has been identified as a Pleistocene fossil, but no extinct Pleistocene species have been described.

Regional Pleistocene Occurrences

Drymarchon corais. Indigo snake: Southeastern region.

Elaphe Fitzinger, 1833: Rat, Corn, and Fox Snakes

These large snakes occur in a wide variety of colors and patterns, some of which are very attractive. They are mainly excellent climbers that mainly feed on small birds and mammals, which they kill by constriction. They occur in much of the eastern and central United States, mainly south of the Great Lakes area, and they also occur in southeastern Canada. Three modern species are known as Pleistocene fossils, but no extinct Pleistocene species have been described.

Regional Pleistocene Occurrences

Elaphe vulpina. Fox snake: Northeastern/Great Lakes, Appalachian, Ozark, and Northern Plains regions.

Elaphe obsoleta. Rat snake: Appalachian, Southeastern, Northern Plains, and Southern Plains regions.

Elaphe guttata. Corn snake: Appalachian, Southeastern, Ozark, Southern Plains, and Rocky Mountain/Great Basin regions.

Farancia Gray, 1842: Mud and Rainbow Snakes

These are very large, shiny snakes with attractive combinations of red, yellow, and black coloration. They are denizens of marsh and swamplands. They occur today as two species, *Farancia abacura,* which is specialized for eating eellike salamanders, and *Farancia erytrogramma,* which is specialized for eating true eels. Both species occur in the southeastern United States. Only *Farancia abacura* has been identified as a Pleistocene fossil, and no extinct Pleistocene species have been described.

Regional Pleistocene Occurrences
Farancia abacura. Mud snake: Southeastern region.

Gyalopion Cope, 1861: Hook-Nosed Snakes

This genus is composed of small, blotched snakes with an upturned snout. These snakes have burrowing habits and they are seldom seen in the daylight hours. They feed on spiders, scorpions, and centipedes. They occur in Texas, Arizona, and southern New Mexico today. Only one of the two modern species of this genus, *Gyalopion canum,* is known as a Pleistocene fossil. No extinct Pleistocene species have been described.

Regional Pleistocene Occurrences
Gyalopion canum. Chihuahuan hook-nosed snake: Southern Plains and Rocky Mountain/Great Basin regions.

Heterodon Latreille, 1801: Hog-Nosed Snakes

These are stout-bodied, medium-sized snakes, normally with a blotched pattern and an upturned snout (Fig. 12B). They are frog- and toad-eating specialists with enlarged rear bladelike teeth for puncturing their inflated prey. They are all able to spread the head and neck widely in a defensive posture and all feign death when molested. They prefer sandy habitats. Hog-nosed snakes are widespread in the United States east of the rockies and occur in parts of southeastern and southwestern Canada today. There are three modern species and all of them are represented by Pleistocene fossils. No extinct Pleistocene species have been described.

Regional Pleistocene Occurrences
Heterodon platirhinos. Eastern hog-nosed snake: Appalachian, Southeastern, Ozark, Northern Plains, and Southern Plains regions.
Heterodon simus. Southern hog-nosed snake: Southeastern region.
Heterodon nasicus. Western hog-nosed snake: Southeastern, Northern Plains, Southern Plains, and Rocky Mountain/Great Basin regions.

Hypsiglena Cope, 1860: Night Snakes

Night snakes are small animals with a blotchy body pattern and with a pair of large brown blotches on the neck. They are rear-fanged snakes that have eyes with vertical pupils. These snakes are relatively ubiquitous in their habitat selection and they feed on salamanders, frogs, lizards, and snakes, which they capture by venomous injections. The single species, *Hypsiglena torquata,* is widespread in many of the southwestern and Pacific Coast states today, and it is also known as a Pleistocene fossil. No extinct Pleistocene species of night snakes have been described.

Regional Pleistocene Occurrences

Hypsiglena torquata. Night snake: Southern Plains, Rocky Mountain/Great Basin, and Pacific Coast regions.

Lampropeltis Fitzinger, 1843: Kingsnakes and Milk Snakes

This group of snakes ranges in size from large to small and in coloration from being rather plain, blotched animals to brilliant coral snake mimics. All of them are constrictors that include snakes in the diet and the larger ones eat rodents and birds. Snakes of this genus eat venomous snakes from time to time and seem to be immune to their poison. This genus occurs in most areas from coast to coast in the United States and also occurs in southeastern Canada. All six modern *Lampropeltis* species are represented by Pleistocene fossils, and an extinct Pleistocene species, *Lampropeltis intermedius* (Fig. 38), has been described that will be discussed in the appropriate regional section.

Regional Pleistocene Occurrences

Lampropeltis triangulum. Milk snake: Northeastern/Great Lakes, Appalachian, Southeastern, Ozark, Northern Plains, Southern Plains, and Rocky Mountain/Great Basin regions.

Lampropeltis getula. Common kingsnake: Appalachian, Southeastern, Ozark, Northern Plains, Southern Plains, Rocky Mountain/Great Basin, and Pacific Coast regions.

Lampropeltis calligaster. Prairie kingsnake: Appalachian, Ozark, Northern Plains, and Southern Plains regions.

Lampropeltis mexicana. Gray-banded kingsnake: Southern Plains region.

Lampropeltis intermedius. Intermediate kingsnake: Rocky Mountain/Great Basin region.

Lampropeltis pyromelana. Sonoran Mountain kingsnake: Rocky Mountain/Great Basin region.

Lampropeltis zonata. California Mountain kingsnake: Rocky Mountain/Great Basin region.

Masticophis Baird and Girard, 1853: Whipsnakes, Coachwhips, and Striped Racers

This genus is composed of large species that are either plain-colored or striped. They are slender, nervous, alert, very fast, nonconstricting forms with very large eyes (Fig. 12C). Their favorite foods are lizards, snakes, and small mammals. These snakes occur from coast to coast in the United States, mainly in the southern half of the country, except for some of the Great Basin and Pacific states, where they have a more northerly distribution. There are four modern species, three of which are known as Pleistocene fossils. No extinct Pleistocene species have been described.

Regional Pleistocene Occurrences

Masticophis flagellum. Coachwhip: Southeastern, Ozark, Rocky Mountain/Great Basin, and Pacific Coast regions.

Masticophis taeniatus. Striped whipsnake: Rocky Mountain/Great Basin and Pacific Coast regions.

Masticophis lateralis. Striped racer: Pacific Coast region.

In addition, "*Masticophis* or *Coluber*" has been reported from the Appalachian region and *Masticophis* sp. has been reported from the Southern Plains region.

Nerodia Baird and Girard, 1853: Water Snakes

Water snakes range from medium to large size and are mainly dull-colored, heavy-bodied aquatic forms. They mainly feed on fishes, although at least one species probably specializes on frogs. Water snakes are widespread east of the Rocky Mountains in the United States in aquatic situations and also occur in southeastern Canada today. Seven modern species occur in North America today, five of which are known in the Pleistocene. No extinct Pleistocene species have been described.

Regional Pleistocene Occurrences

Nerodia sipedon. Northern water snake: Northeastern/Great Lakes, Appalachian, Ozark, and Northern Plains regions.

Nerodia erythrogaster. Plain-bellied water snake: Northeastern/Great Lakes, Appalachian, Southeastern, and Southern Plains regions.

Nerodia fasciata. Southern water snake: Southeastern and Southern Plains regions.

Nerodia cyclopion. Green water snake: Southeastern region.

Nerodia taxispilota. Brown water snake: Southeastern region.

In addition, *Nerodia* sp. has been recorded from the Rocky Mountain/Great Basin region.

Opheodrys Fitzinger, 1843: Green Snakes

These are small, slender, bright green snakes. One species, *Opheodrys aestivus*, prefers bushes and vines, whereas the other, *Opheodrys vernalis*, prefers grassy habitats. Insects are the preferred food for both species. The genus is widespread throughout the United States today except for the Far West and also occurs in parts of southeastern and south-central Canada. Both species are known as Pleistocene fossils, but no extinct Pleistocene species have been described.

Regional Pleistocene Occurrences

Opheodrys vernalis. Smooth green snake: Northeastern/Great Lakes, Appalachian, and Southeastern regions.

Opheodrys aestivus. Rough green snake: Appalachian, Southeastern, and Southern Plains regions.

Phyllorhynchus Stejneger, 1890: Leaf-Nosed Snakes

These are small, spotted or saddled snakes with a modified broad rostrum that is used for burrowing. They feed mainly on lizards and their eggs. In the area of this book, these snakes occur in southern Arizona, Nevada, and California. There are two modern species, one of which, *Phyllorhynchus decurtatus*, is represented by Pleistocene fossils. No extinct Pleistocene species have been described.

Regional Pleistocene Occurrences

Phyllorhynchus decurtatus. Spotted leaf-nosed snake: Rocky Mountain/Great Basin region.

Pituophis Holbrook, 1842: Bullsnakes, Pine and Gopher Snakes

These large snakes are mainly blotched, but a subspecies and some individuals may be plain-colored. They are large constricting snakes that hiss loudly when confronted or molested by other large animals. Their food consists largely of small mammals. The genus is represented by one highly variable species, *Pituophis melanoleucus*, that occurs in the southeastern United States, and then over almost the entire central and western portions. It also occurs in southwestern Canada. The modern species is widespread as a Pleistocene fossil. No extinct Pleistocene species of this genus have been described.

Regional Pleistocene Occurrences

Pituophis melanoleucus. Bullsnakes, pine and gopher snakes: Appalachian, Southeastern, Ozark, Northern Plains, Southern Plains, Rocky Mountain/Great Basin, and Pacific Coast regions.

Regina Baird and Girard, 1853: Crayfish Snakes

These are small, often striped or occasionally rather plain-colored aquatic snakes. They inhabit small streams and prefer crayfish as the main part of their diet. The genus is widespread in the eastern and central United States and also occurs in southeastern Canada. There are four modern species of crayfish snakes, three of which have been identified as Pleistocene fossils. An extinct Pleistocene species, *Regina intermedia*, has been described that will be discussed in the appropriate regional account.

Regional Pleistocene Occurrences

Regina septemvittata. Queen snake: Northeastern/Great Lakes and Appalachian regions.

Regina alleni. Striped crayfish snake: Southeastern region.

Regina intermedia. Intermediate crayfish snake: Southeastern region.

Regina grahamii. Graham's crayfish snake: Northern Plains region.

Rhadinaea Cope, 1863: Pine Woods Snakes

These small, plain-colored snakes occur mainly in the American tropics, where there are about forty species. Only one species, *Rhadinaea flavilata*, occurs in the area covered by this book. This secretive snake of the pine woods eats mainly small frogs and lizards. It has a narrow range from the coastal region of North Carolina to eastern Louisiana and it occurs throughout most of Florida. This snake has also been recorded as a Pleistocene fossil, but no extinct Pleistocene species have been described.

Regional Pleistocene Occurrences

Rhadinaea flavilata. Pine woods snake: Appalachian and Southeastern regions.

Rhinocheilus Baird and Girard, 1853: Long-Nosed Snakes

Long-nosed snakes are medium-sized forms that have saddlelike red and black blotches along the side of the body. These snakes occur in desert and shrubby areas, where they are active at night but spend the daylight hours underground or under some object. Lizards are the preferred food item. Long-nosed snakes are widespread in the southwestern United States. Only one species, *Rhinocheilus lecontei*, occurs in the modern fauna, and this species is also recorded as a Pleistocene fossil. No extinct Pleistocene species have been described.

Regional Pleistocene Occurrences

Rhinocheilus lecontei. Long-nosed snake: Southern Plains, Rocky Mountain/Great Basin, and Pacific Coast regions.

Salvadora Baird and Girard, 1853: Patch-Nosed Snakes

These are medium-sized snakes with a mid-dorsal stripe and a snout that is modified for digging in sand. They are diurnal terrestrial snakes that feed mainly on lizards and small snakes and occasionally consume reptile eggs. In the area of this book, they are found in the southwestern United States. This genus consists of three modern species. The genus has been identified as a Pleistocene fossil, but no specific determinations have been made. No extinct Pleistocene species of this snake have been described.

Regional Pleistocene Occurrences

Salvadora sp. Patch-nosed snake: Rocky Mountain/Great Basin region.

Stilosoma Brown, 1890: Short-Tailed Snakes

These are small blotched snakes with a very slender body and a short tail. They are burrowers that prefer dry pine woodlands and they feed on small snakes and lizards, which they kill by constriction. They occur only in peninsular Florida today. The modern species, *Stilosoma extenuatum*, is known as a Pleistocene fossil, but no extinct Pleistocene species have been described.

Regional Pleistocene Occurrences

Stilosoma extenuatum. Short-tailed snake: Southeastern region.

Sonora Baird and Girard, 1853: Ground Snakes

These are small, plain-colored, secretive snakes that feed on small invertebrates. They occur mainly in the plains and southwestern areas of the United States. This genus consists of a single variable species, *Sonora semiannulata*, that has also been identified as a Pleistocene fossil. No extinct Pleistocene species of *Sonora* have been described.

Regional Pleistocene Occurrences

Sonora *semiannulata*. Ground snake: Northern Plains, Southern Plains, and Rocky Mountain/Great Basin regions.

Storeria Baird and Girard, 1853: Brown and Red-Bellied Snakes

This genus consists of small snakes that are plain-colored above but may have gray, purplish, red, or yellowish bellies. There are two species, both of which are specialized for eating slugs and earthworms. The genus occurs widely in eastern and central United States as well as in southeastern and south-central Canada today. Both modern species have been identified as Pleistocene fossils, but no extinct Pleistocene fossils have been described.

Regional Pleistocene Occurrences

Storeia dekayi. DeKay's brown snake: Appalachian, Southeastern, Northern Plains, and Southern Plains regions.

Storeria occipitomaculata. Red-bellied snake: Appalachian region.

In additiona, *Storeria* sp. has been recorded from the Ozark region.

Tantilla Baird and Girard, 1853: Black-Headed Snakes

These are very small snakes with plain-colored bodies and black or brown heads. Ground-dwelling and secretive, these snakes feed upon small invertebrates. They have enlarged teeth in the rear of the mouth and they have poison that helps to subdue their prey. This genus occurs widely in the southeastern and southwestern United States today and is composed of at least 11 modern species, only one of which, *Tantilla relicta*, has been identified as a Pleistocene fossil. No extinct Pleistocene species have been described.

Regional Pleistocene Occurrences

Tantilla relicta. Florida crowned snake: Southeastern region.

In addition, *Tantilla* sp. has been recorded from the Southeastern, Northern Plains, Southern Plains, Rocky Mountain/Great Basin, and Pacific Coast regions.

Thamnophis Fitzinger, 1843: Garter and Ribbon Snakes

These are small to medium-sized, normally striped snakes that are ubiquitous in their choice of habitats. They feed on a wide variety of soft-bodied invertebrates and cold-blooded vertebrates, sometimes eating shrews or mice. Because of their prolific reproductive habits, some species may be the most frequently found snakes in any given region. This large genus is widespread in almost all of the United States and reaches far north into Canada. In the area of this book, at least thirteen modern species occur, nine of which have been identified as Pleistocene fossils. No extinct Pleistocene species of *Thamnophis* have been described.

Regional Pleistocene Occurrences

Thamnophis sirtalis. Common garter snake: Northeastern/Great Lakes, Appalachian, Southeastern, Ozark, Northern Plains, Southern Plains, Rocky Mountain/Great Basin, and Pacific Coast regions.

Thamnophis brachystoma. Short-headed garter snake: Appalachian region.

Thamnophis sauritus. Eastern ribbon snake: Appalachian and Southeastern regions.

Thamnophis proximus. Western ribbon snake: Ozark, Northern Plains, Southern Plains, Rocky Mountain/Great Basin, and Pacific Coast regions.

Thamnophis radix. Plains garter snake: Northern Plains region.

Thamnophis marcianus. Checkered garter snake: Southern Plains region.

Thamnophis cyrtopsis. Black-necked garter snake: Rocky Mountain/Great Basin and Pacific Coast regions.

Thamnophis elegans. Western terrestrial garter snake: Rocky Mountain/Great Basin and Pacific Coast regions.

Thamnophis couchii. Sierra garter snake: Pacific Coast region.

Trimorphodon Cope, 1861: Lyre Snakes

These medium-sized, grayish, blotched snakes live in rocky areas where they immobilize their prey by injecting venom by means of enlarged, grooved teeth in the back of the mouth. They feed on lizards, birds, and small mammals. In the area of this book, lyre snakes occur in the far southwestern United States today. The genus consists of a single species, *Trimorphodon biscutatus*, which has also been reported as a Pleistocene fossil. No extinct Pleistocene species of the genus have been described.

Regional Pleistocene Occurrences

Trimorphodon biscutatus. Western lyre snake: Rocky Mountain/Great Basin region.

Tropidoclonion Cope, 1860: Lined Snakes

These are small, brownish, lightly striped snakes. They are secretive in their habits, and earthworms are their favorite food. They are widespread in the central "prairie" regions of the United States today. The genus consists of a single species, *Tropidoclonion lineatum*, which also has been recorded as a Pleistocene fossil. No extinct Pleistocene species have been described.

Regional Pleistocene Occurrences

Tropidoclonion lineatum: Northern Plains and Southern Plains regions.

Virginia Baird and Girard, 1853: Earth Snakes

These are very small, nondescript, plain-colored, highly secretive snakes that feed upon earthworms and other soft-bodied invertebrates. Earth snakes occur chiefly in the southeastern United States west into Texas today. Two species occur in the genus; both have been recorded as Pleistocene fossils, but no extinct Pleistocene species have been described.

Regional Pleistocene Occurrences
Virginia striatula. Rough earth snake: Appalachian and Southern Plains regions.
Virginia valeriae. Smooth earth snake: Appalachian region.
In addition, *Virginia* sp. has been reported from the Southeastern region.

Family Elapidae Boie, 1827

Elapids (Figs. 13A, 28C)

This largely tropicolitan family consists of snakes with hollow, immovable fangs and extremely potent venom. The family is represented in the area of this book by two genera, *Micruroides* and *Micrurus*, but only the latter species has been recorded as a Pleistocene fossil. No extinct Pleistocene elapid genera have been described.

Osteological characters include a skull that is very similar to the Colubridae, except for the anterior, fixed, tubular fangs. The vertebrae (Fig. 28C) are also very similar to the Colubridae. In all but a few rare elapids, hypapophyses are present throughout the entire trunk region of the vertebral column.

Micrurus Wagler, 1824: American Coral Snakes

These small, brilliantly colored snakes have a body ringed with black, yellow, and red bands. They are secretive animals that feed mainly on small lizards and

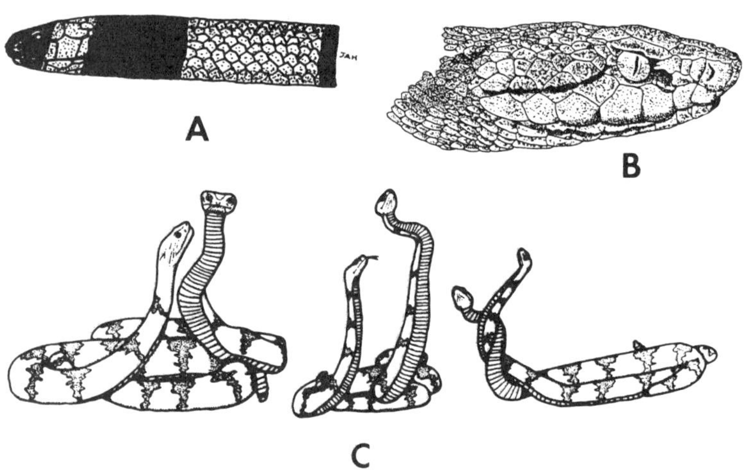

FIGURE 13. Poisonous snakes. (A) *Micrurus fulvius* (Elapidae). (B) *Agkistrodon contortrix* (Viperidae). (C) *Crotalus lepidus* (Viperidae) males in ritual combat poses. These species occurred in the Pleistocene of North America. [A, drawing by the author; B, drawing by C. Smith, permission of *Herpetological Review;* C from Carpenter, Gillingham, and Murphy, 1976, permission of *Copeia*]

snakes. They occur in the southeastern United States today, eastward into western central Texas. Only a single species, *Micrurus fulvius*, occurs today in the area of this book, and this species has also been identified as a Pleistocene fossil. No extinct Pleistocene species of this genus have been described.

Regional Pleistocene Occurrences

Micrurus fulvius. North American coral snake: Southeastern and Southern Plains regions.

Family Viperidae

Vipers, Adders, and Pit Vipers (Figs. 13B,C, 28D–F, 39)

This family is very widespread in the temperate and tropical regions of the world. These snakes have large, curved, hollow fangs that occur on a movable maxillary bone and a very deadly venom. Three genera occur in the modern fauna of North America and all of them have been recorded as Pleistocene fossils. No extinct Pleistocene genera have been described.

Osteological characters include an ectopterygoid that is as long as the quadrate and a quadrate that slants posteriorly. The vertebrae (Fig. 28D–F) have hypapophyses (see Fig. 26A) that are very thick and occur throughout the trunk region of the vertebral column.

Agkistrodon Beauvois, 1799: Copperheads and Cottonmouths

These large, thick-bodied snakes are either dull-colored or have an attractive pattern of brownish or reddish brown bands on a lighter body color. They are either terrestrial animals (e.g., *Agkistrodon contortrix,* Fig. 13B) that mainly feed on frogs and rodents, or aquatic ones (e.g., *Agkistrodon piscivorus*) that feed largely on fish. Only the species just mentioned occur in the area of this book. These animals occur mainly in the eastern and southeastern United States west into Kansas and Texas. Both modern species have been recorded as Pleistocene fossils, but no extinct Pleistocene species have been described.

Regional Pleistocene Occurrences

Agkistrodon contortrix. Copperhead: Appalachian, Northern Plains, and Southern Plains regions.

Agkistrodon piscivorus. Cottonmouth: Appalachian, Southeastern and Southern Plains regions.

Crotalus Linnaeus, 1758: Rattlesnakes

These snakes are generally large, blotched, banded, or sometimes spotted, with a variety of color patterns, and a rattle on the end of the tail. Most of them feed on small mammals or birds, which they kill with their powerful venom. They occur in a wide variety of habitats, ranging from damp woodlands to deserts. This large genus occurs widely throughout the United States, except for most of the Great Lakes region, and in southeastern and southwestern Canada today. Thirteen species occur in the area of this book, eight of which have been identified as Pleistocene fossils. An extinct Pleistocene species, *Crotalus potterensis* (Fig. 39), has been described and will be addressed in the appropriate regional section.

Regional Pleistocene Occurrences

Crotalus horridus. Timber rattlesnake: Northeastern/Great Lakes, Appalachian, Ozark, Northern Plains, and Southern Plains regions.
Crotalus adamanteus. Eastern diamondback rattlesnake: Southeastern region.
Crotalus viridis. Western rattlesnake: Northern Plains, Southern Plains, Rocky Mountain/Great Basin, and Pacific Coast regions.
Crotalus atrox. Western diamondback rattlesnake: Southern Plains, Rocky Mountain/Great Basin, and Pacific Coast regions.
Crotalus cerastes. Sidewinder: Rocky Mountain/Great Basin region.
Crotalus lepidus. Rock rattlesnake: Rocky Mountain/Great Basin region.
Crotalus mitchellii. Speckled rattlesnake: Rocky Mountain/Great Basin region.
Crotalus scutulatus. Mojave rattlesnake: Rocky Mountain/Great Basin region.
Crotalus potterensis. Potter Cave rattlesnake: Pacific Coast region.

Sistrurus Garman, 1883: Massasaugas and Pygmy Rattlesnakes

These are small to medium-sized rattlesnakes with blotched patterns. They occur from swampy to dry habitats, with some populations occurring in both situations, depending upon the time of the year. Frogs, snakes, and mice are favored prey items and they are killed by venomous injections. These snakes occur in eastern, southeastern, and plains regions of the United States, as well as in southeastern Canada today. There are two modern species of *Sistrurus*, both of which have been recorded as Pleistocene fossils. No extinct Pleistocene species have been described.

Regional Pleistocene Occurrences

Sistrurus catenatus. Massasauga: Appalachian, Northern Plains and Southern Plains regions.
Sistrurus miliarius. Pygmy rattlesnake: Southeastern region.

Order Crocodilia Gmelin, 1788: Crocodilians

Two families of crocodilians, the Alligatoridae (Fig. 14) and the Crocodylidae, occur in the southeastern United States today, but only the Alligatoridae has been definitely identified as a Pleistocene fossil.

Family Alligatoridae Gray, 1844
Alligators and Caimans

Only a single species, *Alligator mississippiensis* (Fig. 14), a very large, dark-colored, blunt-nosed crocodilian of shallow lakes and swamps, occurs in the area of this book in North America today. Alligators feed on a wide variety of vertebrate food, including fishes, turtles, snakes, birds, and mammals. They

FIGURE 14. A representative of the North American crocodilian family Alligatoridae. *Alligator mississippiensis*. This species occurs in the Pleistocene of North America. [Drawing by T. R. Johnson; permission of *Herpetological Review*]

build large, moundlike nests on land and exhibit parental care of the young, as in other crocodilian species. No extinct Pleistocene alligatorid species have been described in North America.

Important osteological characters of the family include a relatively short, broad rostrum and the fact that the fourth mandibular tooth is not visible when the jaw is closed. The osteoderms that occur on the backs of alligators have a very characteristic medial ridge.

Alligator Cuvier, 1807: Alligators

Alligators today occur mainly in the southeastern United States down into the Florida keys, ranging westward into Texas and northward into Oklahoma.

Regional Pleistocene Occurrences

Alligator mississippiensis. American alligator: Southeastern, Northern Plains, and Southern Plains regions.

In addition, alligators have tentatively been identified in Missouri caves. These finds will be addressed in the appropriate regional accounts.

5

IDENTIFICATION OF NORTH AMERICAN PLEISTOCENE HERPETOLOGICAL FOSSILS

Most North American Pleistocene herpetological fossils occur as scattered postcranial bones or parts of bones or as broken pieces of skulls and jaws. Occasionally, more complete fossils may be found, such as complete turtle shells (Figs. 42, 43) or even skulls of various taxa, but these occurrences are rare. Individual bones that are used for the identification and study of amphibians and reptiles generally must satisfy two criteria: (1) they must be of common occurrence, and (2) they must have useful characters for the identification of the animals that bore them.

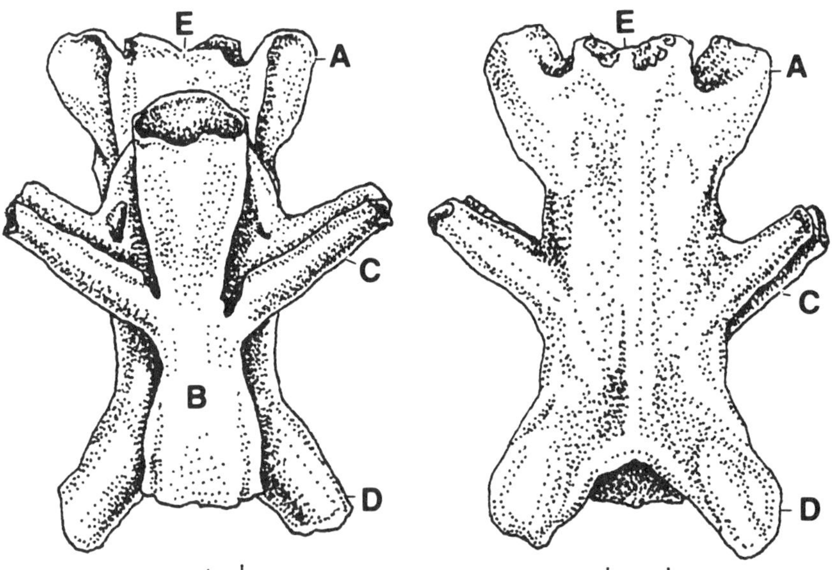

FIGURE 15. Trunk vertebra of *Ambystoma jeffersonianum* complex (Ambystomatidae), Wisconsinan New Trout Cave, Pendleton County, West Virginia, indicating taxonomically important structures. (A) postzygapophysis; (B) centrum; (C) rib-bearing processes; (D) prezygapophysis; (E) posterior

Since many delicate skull bones break up during their transportation to the fossil site, the sturdier postcranial bones (and sometimes dentary bones) are the ones that normally withstand the test of time. But these bones differ a great deal in their taxonomic usefulness. Over the past forty years, since the surge of interest in the study of fossil amphibians and reptiles of modern orders began, certain elements have been relied upon for identification more than others.

In general, bones that bear teeth, or that have a good many muscle attachments, or that have a quite complex structure are the most useful in making herpetological identifications. In this chapter we will look at some of the bones widely used in the identification of North American Pleistocene herpetofauna, discuss some important characters of each of these elements, and mention some of the difficulties involved in the identification of fossil taxa on the basis of single elements.

Salamanders

The most commonly used elements to identify Pleistocene salamanders are trunk vertebrae (Figs. 15–17). Other elements might possibly be more diagnostic, but salamander skeletons are so cartilaginous, and many of the bones so fragile, that salamander vertebrae are often the only elements that turn up at fossil sites. A classic paper on the use of trunk vertebrae in the identification of fossil ambystomatid salamanders is that of Tihen (1958).

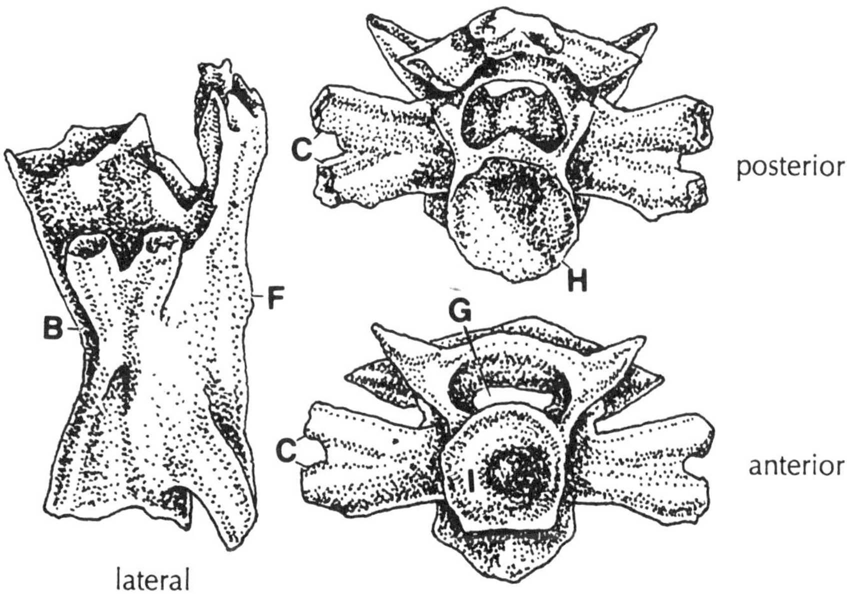

border of neural arch; (F) neural spine; (G) neural canal; (H) posterior cotyle; and (I) anterior cotyle. [Modified from Holman and Grady, 1987]

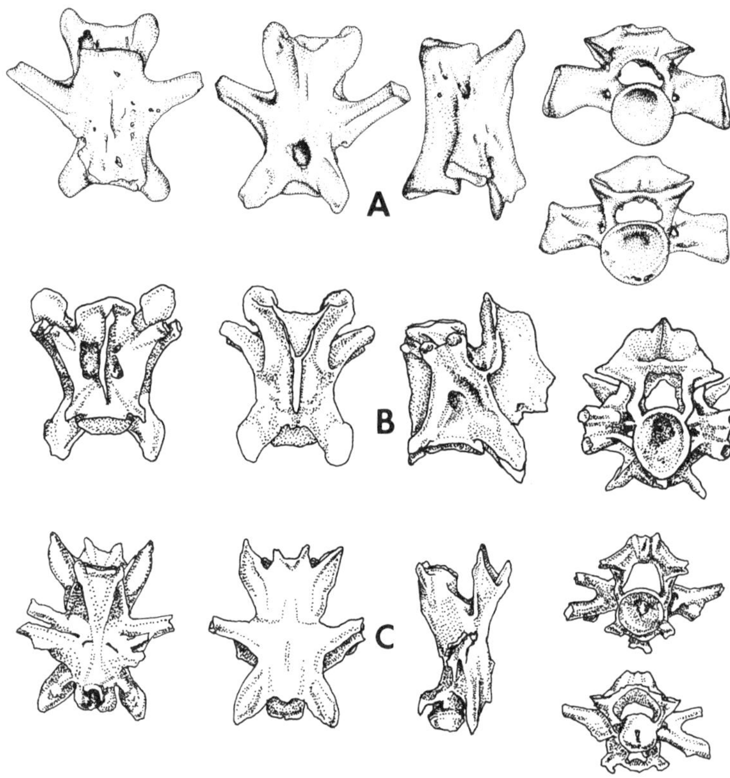

FIGURE 16. (A) simple, (B) complex, and (C) intermediately complex North American salamander trunk vertebrae. (A) *Cryptobranchus guildayi* (Cryptobranchidae), Irvingtonian Trout Cave site, Pendleton County, West Virginia. (B) *Siren lacertina* (Sirenidae) (modern skeleton). (C) *Desmognathus ochrophaeus* (Plethodontidae), Wisconsinan New Trout Cave Site of Pendleton County, West Virginia. [A from Holman, 1982; B from Holman and Voorhies, 1985; C from Holman and Grady, 1987]

Salamander trunk vertebrae range from those that have rather simple structure with low neural spines (e.g., Cryptobranchidae, Fig. 16A; and Ambystomatidae, Figs. 15, 17E) to those with a complicated structure with well-developed and elaborated neural spines (e.g, Sirenidae, Fig. 16B). Others are rather intermediate in their structural complexity, such as those of some of the Plethodontidae (Fig. 16C). Some salamanders have trunk vertebrae with ossified caps on their cotyla, making these vertebrae appear opisthocoelous (Figs. 16C, 17F).

Problems in the identification of individual salamander vertebrae involve (1) individual variations; (2) lack of specific differences in trunk vertebrae (e.g., species of *Plethodon*); (3) intercolumnar variations in trunk vertebrae; (4) ontogenetic variations; and (5) loss of elements fused to the vertebrae in life such as the cotylar caps, which make the living animals appear to be opisthocoelous, or the loss of co-ossified portions of the neural spine.

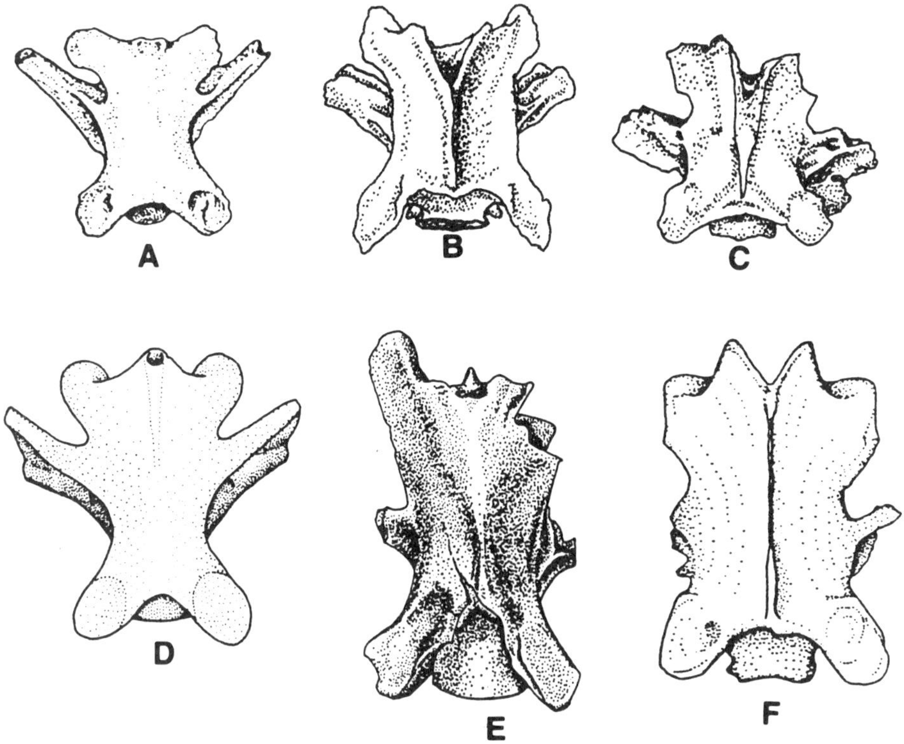

FIGURE 17. Representative North American salamander trunk vertebrae in dorsal view. (A) *Cryptobranchus alleganiensis* (Cryptobranchidae), Wisconsinan Bell Cave site, Colbert County, Alabama. (B) *Siren lacertina* (Sirenidae) (modern skeleton). (C) *Notophthalmus* sp. (Salamandridae), Irvingtonian Fyllan Cave site, Austin County, Texas. (D) *Necturus maculosus* (Proteidae) (modern skeleton). (E) *Ambystoma texanum* (Ambystomatidae), Wisconsinan Howard Ranch fauna, Hardeman County, Texas. (F) *Desmognathus ochrophaeus* (Plethodontidae), Wisconsinan Bell Cave site, Colbert County, Alabama. [A from Holman, Bell, and Lamb, 1990; B from Holman and Voorhies, 1985; C from Holman and Winkler, 1987; D from Naylor, 1978; E from Holman, 1964; F from Holman, Bell, and Lamb, 1990]

Anurans

The anuran ilium appears to be the single best element upon which to base identifications when only disarticulated individual bones are available as fossils. The osteological structure of the ilium (and to a lesser extent, the sacrum) reflects the mode of locomotion of the animals. Long-leaping anurans, hopping anurans, running or dashing anurans, walking anurans, climbing anurans, and burrowing anurans all have different ilial structures, especially in the posterior part of the bone where the important ligaments and muscles attach. A classic paper on the use of the ilium in the identification of fossil Bufonidae is that of Tihen (1962).

Anuran ilia may be divided into two groups, those with an ilial blade or crest, often called a vexillum in European termnology (e.g., Ranidae, Fig. 18A,B) and those that lack such a structure (e.g., Bufonidae, Fig. 19C,D; and Hylidae, Fig. 19E,F). In taxa that have a crest, the relative development of the crest (bladelike or only ridgelike) and the size and shape of the crest are important characters. Sometimes the posterior part of the crest is elaborated with well-developed muscle attachments that are taxonomically useful.

In anuran taxa that lack an ilial blade or crest, the prominence that occurs on the dorsal part of the ilium near the acetabular facet is important. This prominence may be obsolete or absent (e.g., Pelobatidae, Fig. 19A), low and long (e.g., some Bufonidae, Fig. 19C), high and narrow (e.g., some Bufonidae, Fig. 19D), triangular (e.g., Microhylidae, Fig. 20F–H), or round and rather laterally directed (e.g., Hylidae, Fig. 19E,F). The shape of the expanded areas above and below the acetabular fossa are also important characters.

Some workers have attempted to use the anuran sacrum as an element upon which to base fossil identifications. The anuran sacrum consists of a single vertebra with laterally directed sacral diapophyses (Fig. 21B). The sacrum may either be procoelous with an anterior cotyle and two posterior condyles (e.g., Bufonidae, Fig. 21A) or diplasiocoelous with an anterior and two posterior

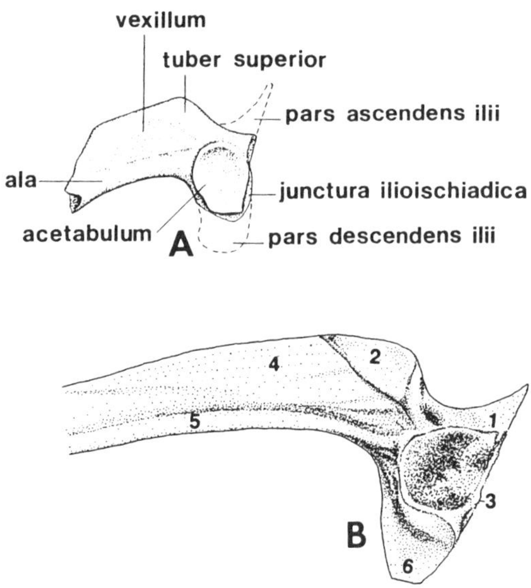

FIGURE 18. Ilia of *Rana* indicating taxonomically important structures. (A) *Rana pipiens* complex (Ranidae) left ilium from Lisco C Blancan fauna of Garden County, Nebraska, illustrating ilial terminology often used in Europe. (B) *Rana clamitans* (Ranidae) left ilium from a modern skeleton illustrating ilial terminology used in North America: 1, dorsal acetabular expansion; 2, dorsal prominence (sometimes called vastus prominence); 3, border of acetabulum; 4, ilial crest or blade; 5, ilial shaft; 6, ventral acetabular expansion. [A from Holman and Schloeder, 1991; B, drawing by D. R. Holman]

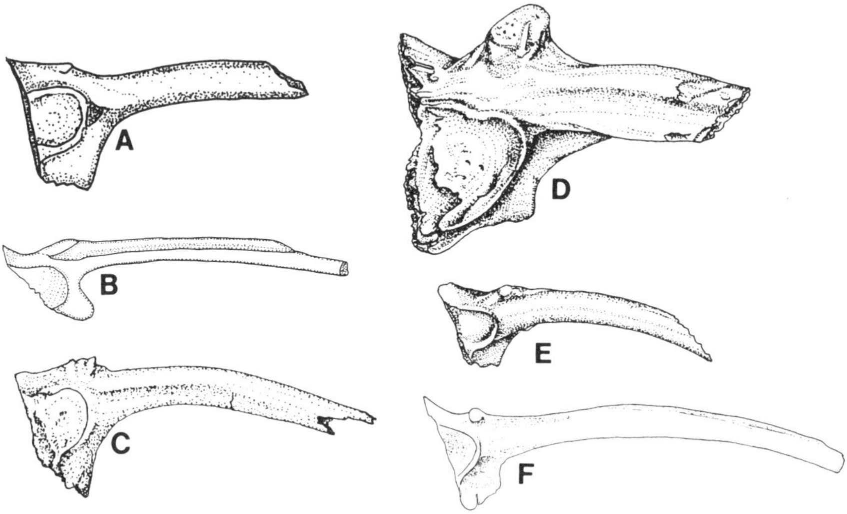

FIGURE 19. Right ilia from representative North American anuran families. (A) *Scaphiopus holbrookii* (Pelobatidae) (undated cave sediments in Florida). (B) *Syrrhophus marnockii* (Leptodactylidae) (modern skeleton). (C) *Bufo americanus* (Bufonidae), Wisconsinan, St. Clair County, Michigan. (D) *Bufo woodhousii woodhousii* (Bufonidae), Illinoian Williams local fauna, Rice County, Kansas. (E) *Acris crepitans* (Hylidae), same locality as D. (F) *Hyla squirella* (Hylidae) (modern skeleton). [A and B, drawings by the author; C from Holman, 1988; D and E from Holman, 1984; F from Lynch, 1965]

condyles (e.g., Ranidae, Fig. 21B). The sacral diapophyses range from cylindrical (e.g., Ranidae, Fig. 21B) to widely expanded (e.g., Pelobatidae, Fig. 21C). Sometimes the sacral vertebra is fused to the urostyle (e.g., Pelobatidae, Fig. 21C).

Convergence of characters in the ilium and sacrum that are based on differences in mode of locomotion occur in families of modern anurans. As an example, some long-leaping leptodactylid frogs, which externally resemble some ranid frogs, also have ilia that resemble ranid frogs; and the same is true with some toadlike leptodactylids and true toads of the family Bufonidae.

Problems that exist in using these elements for the identification of Pleistocene anuran taxa mainly deal with ontogenetic and individual variations that occur. For instance, a *Bufo* species might have an ilial prominence that may generally be described as long and low; but that condition might range from twice as long as high to almost as long as high. In European *Bufo bufo* there are two distinctive morphs in the ilial prominence, one that is low and rounded and a less common one that is irregular and somewhat bladelike (Holman, 1989). This dimorphism may occur in North American *Bufo* as well. Many other individual variations in the ilium and sacrum exist, so that it is necessary to have adequate series of modern comparative skeletons in order to make firm identifications.

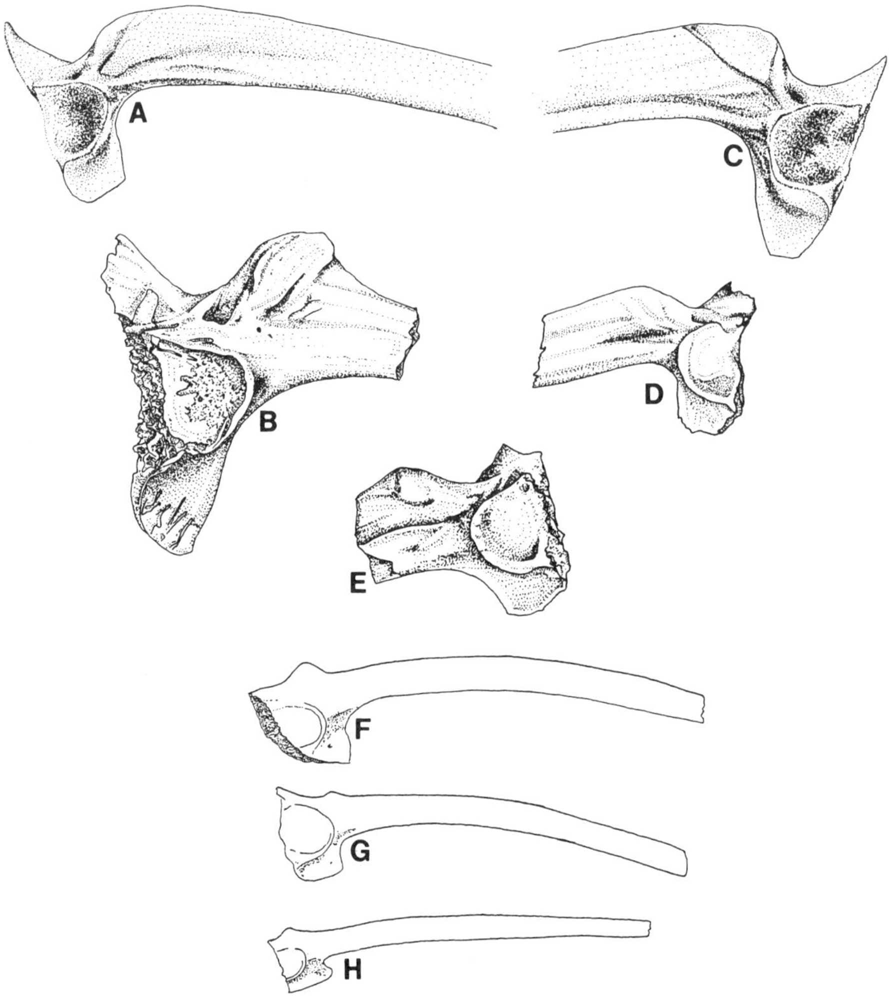

Figure 20. Ilia from representative North American anuran families. (A) *Rana areolata* (Ranidae) right ilium (modern skeleton). (B) *Rana catesbeiana* (Ranidae) right ilium, Illinoian Williams fauna, Rice County, Kansas. (C) *Rana clamitans* (Ranidae) left ilium (modern skeleton). (D) *Rana pipiens* complex (Ranidae) left ilium, same locality as B. (E) *Rana sylvatica* (Ranidae) left ilium, Illinoian Duck Creek fauna, Ellis County, Kansas. (F) *Hypopachus variolosus* (Microhylidae) right ilium (modern skeleton). (G) *Gastrophryne carolinensis* (Microhylidae) (modern skeleton). (H) *Gastrophryne olivacea* (Microhylidae) (modern skeleton). [A and C, drawings by D. R. Holman; B, D, and E from Holman, 1984; F–H from Holman, 1963]

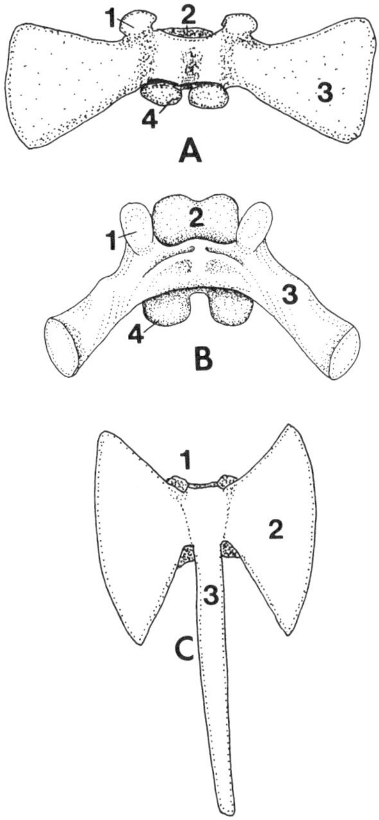

FIGURE 21. Sacra of North American anurans. (A) generalized sacrum of *Bufo* (Bufonidae): (1) prezygapophysis; (2) anteriorly directed cotyle; (3) expanded sacral diapophyses; (4) posteriorly directed condyles. (B) generalized sacrum of *Rana* (Ranidae): (1) prezygapophysis; (2) anteriorly directed condyle; (3) cylindrical sacral diapophyses; (4) posteriorly directed condyles. (C) Generalized sacrum of *Scaphiopus* (Pelobatidae), note sacrum is fused to urostyle: (1) prezygapophysis; (2) greatly expanded sacral diapophyses; (3) urostyle fused to sacrum. [A and C, drawings by the author; B from Hallock, Holman, and Warren, 1990]

Turtles

The most common fossil elements of turtles are individual shell bones (Fig. 22A) or parts of these bones. Occasionally entire turtle shells occur as Pleistocene fossils (Figs. 42, 43), but these are fortuitous exceptions to the rule. Complete skulls are even more exceptional finds. Contrary to the opinions of a few workers, Pleistocene turtle species may often be identified with confidence from single shell bones such as nuchals, hyoplastra, and hypoplastra, or combinations of these elements. This is due in large part to the fact that the

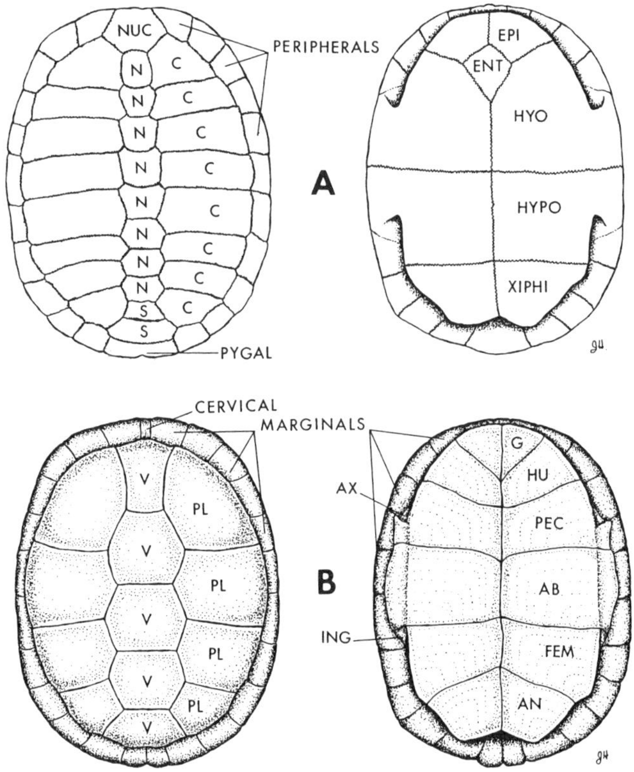

FIGURE 22. Bones (A) and overlying scutes (B) of a generalized emydid turtle. Key for A: NUC = nuchal, N = neural, S=suprapygal, C = costal, EPI = epiplastron, ENT = entoplastron, HYO = hyoplastron, HYPO = hypoplastron, XIPHI = xiphiplastron. Key for B: V = vertebral, PL = pleural, AX = axial, ING = inguinal, G = gular, HU = humeral, PEC = pectoral, AB = abdominal, FEM = femoral, AN = anal. [Drawings by J. H. Harding in Holman and Harding, 1977]

dermal bones of the turtle shell (Fig. 22A) not only may have specifically variable shapes (Fig. 23) but are laced by a pattern of (often specifically distinct) lines produced by the impressions of the overlying epidermal scutes (Fig. 22B).

Over the years, the nuchal bone (Fig. 23) has been recognized as the best turtle shell element to use for specific identifications. Nuchal bones have been especially useful in the study of Pleistocene turtles in North America. Preston (1979) made a definitive study of these elements in a classic paper on Pleistocene turtles of the central plains region of the United States.

The nuchal bone (Fig. 23) is the most anterior of the median series of unpaired bones in the turtle carapace (Fig. 22A). It may be very short and wide as in the Chelydridae (Fig. 23B) or about as long as wide as in most of the Emydidae (Fig. 23F–H). The lines left by the impressions of the epidermal scutes are important taxonomic characters.

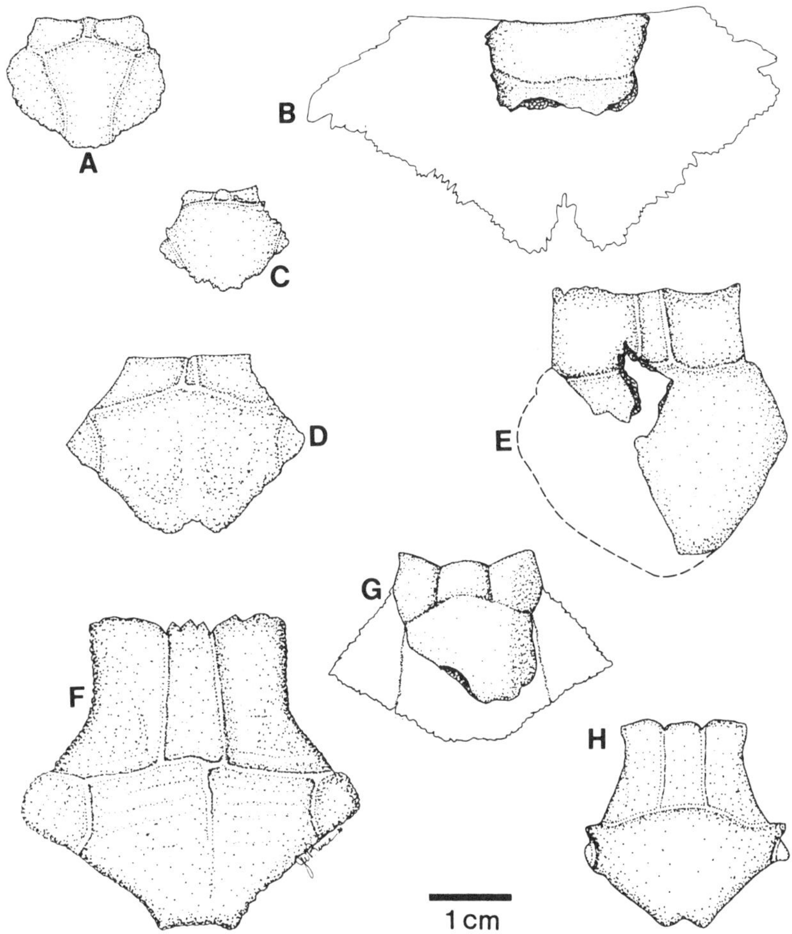

FIGURE 23. Nuchal bones representing various North American taxa (Pleistocene sites, southwestern Kansas). (A) *Sternotherus odoratus* (Kinosternidae). (B) *Chelydra serpentina* (Chelydridae). (C) *Kinosternon flavescens* (Kinosternidae). (D) *Terrapene ornata* (Emydidae). (E) *Emydoidea blandingii* (Emydidae). (F) *Trachemys scripta* (Emydidae). (G) *Graptemys geographica* or *kohni* (Emydidae). (H) *Chrysemys picta* (Emydidae). [From Preston, 1979]

The presence or absence of a cervical (nuchal) scute impression is an important lead to specific identification. For instance, within the tortoise genus *Geochelone*, some species have cervical scutes and others do not. If the cervical scute is present, its shape may be important. The tortoise genus *Gopherus* may be distinguished from those species of the tortoise genus *Geochelone* with a cervical scute on the basis that the cervical scute impression is wider than long in *Gopherus* but is longer than wide in *Geochelone*.

Other characters that involve the cervical scute area of the nuchal bone in turtles include whether that area is elevated or not, whether it extends beyond the anterior margin of the carapace or not, and whether it is blunt or pointed. In some cases the anterior edge of the nuchal bone is serrated (e.g., *Trachemys,* Fig. 23F) and in other cases it is smooth (e.g., *Terrapene,* Fig. 23D).

Sculpturing patterns on the nuchal bone as well as on other turtle carapacial bones (Fig. 24A,B) are important taxonomic characters. It has been shown that species of Pleistocene *Pseudemys* may be distinguished on the basis of the presence or absence of these patterns on certain bones (Holman, 1985a). In some genera there may be no sculpturing patterns (e.g., *Chrysemys*) or heavy sculpturing patterns (e.g., *Pseudemys*). The sculpturing may consist of pits and ridges (e.g., *Apalone,* Fig. 24B) or of ridges and valleys (*Pseudemys*, Fig. 24A).

Of lesser importance, but still useful in the identification of species of Pleistocene turtles, are the hyoplastral and hypoplastral bones (Fig. 24B). The shape and thickness of these bones as well as the structure of their dorsally projecting processes are important. Moreover, the pattern of epidermal scute impression lines is taxonomically important, especially those of the axillary and inguinal scutes.

Variational problems do occur, and these usually involve individual and ontogenetic variations in the size, shape, and other characteristics of turtle shell bones. Occasionally, congenital deformities of shell bones and congenital changes in scute patterns occur in fossil as well as modern turtles. These are usually easy to detect by the trained eye. Nevertheless, it is necessary to have adequate series of comparative skeletons of modern turtle species available to be able to make firm specific identifications of corresponding fossils.

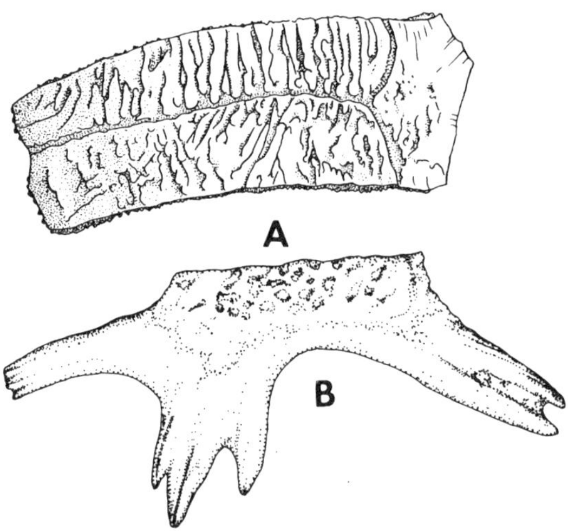

FIGURE 24. North American turtle shell bones. Costal bone (A) of *Pseudemys nelsoni* (Emydidae), Wisconsinan Little Salt Spring site, Sarasota County, Florida, showing ridge and valley sculpturing pattern. Left hypoplastral bone (B) of *Apalone ferox* (Trionychidae) (modern skeleton) showing pit and ridge sculpturing. [A from Holman and Clausen, 1984; B from Holman, 1982]

Lizards and Amphisbaenians

The classic work on lizard fossils is that of Estes (1983). Here many diagnostic bones used to identify fossil lizards are figured. The bones most commonly used to identify Pleistocene lizard taxa are dentary bones with teeth. The dentary bones themselves may be relatively long and narrow (e.g., *Anolis*, Fig. 25A) or short and stout (e.g., *Phrynosoma*, Fig. 25C). The number of nutrient foramina

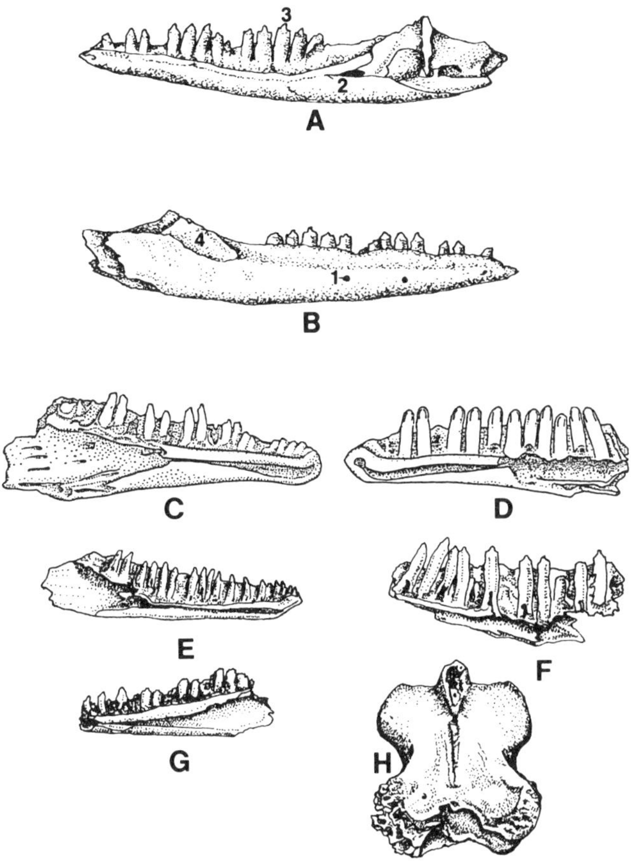

FIGURE 25. Representative North American Pleistocene lizard elements. (A) Medial and (B) lateral view of right dentary of *Anolis carolinensis* (Polychridae), late Wisconsinan Ladds fauna, Bartow County, Georgia: (1) nutrient foramen; (2) Meckelian groove mainly closed; (3) tricuspid, unstriated teeth, (4) splenial groove. (C) Left dentary of *Phrynosoma cornutum* (Phrynosomatidae), Irvingtonian Nash local fauna, Meade County, Kansas. (D) Right dentary of *Eumeces obsoletus,* same locality as C. (E) Left dentary of *Eumeces laticeps* (Scincidae), Wisconsinan New Trout Cave fauna, Pendleton County, West Virginia. (F) Right dentary of *Sceloporus undulatus* (Phrynosomatidae), same locality as E. (G) Right dentary of *Crotaphytus* cf. Crotaphytus *reticulatus* (Crotaphytidae), Irvingtonian Fyllan Cave site, Austin County, Texas. (H) Vertebra of *Ophisaurus* sp. (Anguidae), same source as G. [A and B from Holman, 1985a; C and D from Holman, 1979; E and F from Holman and Grady, 1987; G and H from Holman and Winkler, 1987]

on the lateral surface of the dentary bone (see Fig. 25A) may be of taxonomic importance. On the medial surface of the lizard dentary bone lies a groove called the Meckelian groove (Fig. 25A). This groove may be widely open (Fig. 25C), narrowly open (Fig. 25A), or closed.

Variations in the dentary teeth of lizards are many. The teeth may have darkly pigmented tips, as in some *Eumeces* species or they may be unpigmented. The tips of lizard teeth may also be striated (e.g., *Eumeces*, Fig. 25D) or unstriated (e.g., *Phrynosoma* Fig. 25C). They may be slender and pointed or relatively thick and blunt. They may be unicuspid, bicuspid, or even tricuspid in some forms. Modern species of lizards often have different patterns of tooth replacement which may be sometimes detected in Pleistocene fossils.

Vertebrae are important elements for the identification of certain elongate, legless lizards and in the limbless Pleistocene amphisbaenians. The vertebrae of glass lizards of the genus *Ophisaurus* are characterized by their simplicity (Fig. 25H). They have smooth ventral surfaces and relatively simple neural spines. The vertebrae of Pleistocene amphisbaenians of the genus *Rhineura* are also very characteristic elements with longitudinal striations on the dorsal surface of their neural arches.

Variational problems in the use of individual lizard bones in the identification of specific taxa may be troublesome. Species of lizards in some genera (e.g., *Sceloporus*) are almost identical in many of their osteological characters, although the living forms may easily be distinguished on the basis of external characteristics.

In some modern species of lizards, teeth with different morphologies occur in the same dentary bone. Thus, small portions of fossil dentary bones may not be adequate for diagnoses. There are also cases in several modern lizard species where tooth morphologies change from the juvenile to the adult stage. There are also the expected individual variations in tooth structure as well as the shape and number of nutrient foramina in the dentary bone itself.

In taxa distinguished on the basis of vertebrae (e.g., *Ophisaurus*), there is not only individual variation to worry about, but there are subtle changes in vertebral shape along the extent of the vertebral column. For all these reasons, then, it is very important to have adequate series of recent comparative skeletons of lizards and amphisbaenians to be able to do a satisfactory job in the identification of Pleistocene taxa.

Snakes

Snakes are the most highly derived animals of all of the modern amphibians and reptiles. The limbs and limb girdles are absent (except for tiny remnants in some very primitive snakes), but the remaining units, skull, vertebral column, and ribs, are exceedingly modified. The bones of the kinetic snake skull are fragile and are infrequently found as fossils. Moreover, many of the snake skull bones that do fossilize are not very diagnostic. The ribs of snakes are often found as fossils, but snake ribs are not very helpful in making specific taxonomic identifications.

For these reasons the vertebrae of snakes (Figs. 26–28) have been highly utilized in the identification of fossil snakes, especially those from the Pleistocene of North America. Auffenberg (1963b) paved the way for the modern study of Pleistocene snakes in North America in his classic work on the fossil snakes of Florida mainly on the basis of vertebral remains.

The integrated vertebral column and rib functional unit of snakes provide them with the ability to have a variety of patterns of motion unexcelled by any other living vertebrate group. All snakes can telescope their bodies, glide, wriggle, climb, burrow to a certain extent, swim, tie themselves in knots, and

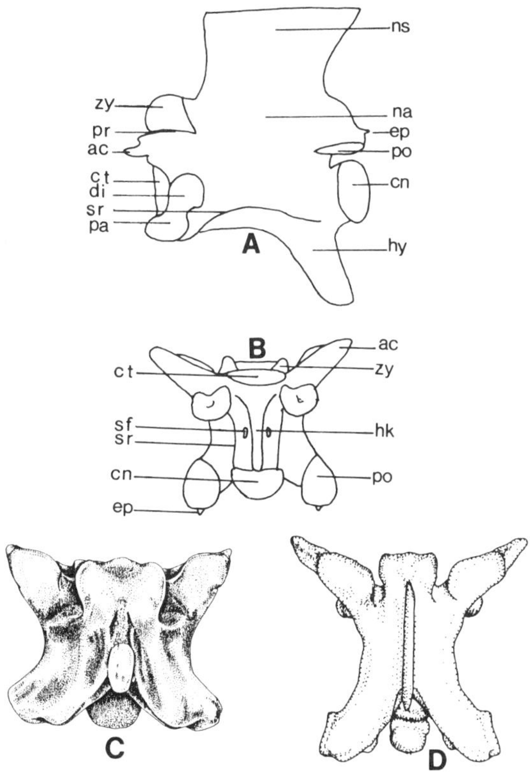

FIGURE 26. Snake trunk vertebrate. (A) Natricine type of colubrid snake trunk vertebra in lateral view. (B) Colubrine type of colubrid trunk vertebra in ventral view. Starting at the top left side of A: ZY = zygosphene; PR = prezygapophyseal articular facet; AC = prezygapophyseal accessory process; CT = cotyle; DI = diapophyseal portion of synapophysis; SR = subcentral ridge; PA = parapophyseal portion of synapophysis; NS = neural spine; NA = neural arch; EP = epizygapophyseal spine; PO = postzygapophyseal articular facet; CN = condyle; HY = hypapophysis; SF = subcentral foramen; HK = hemal keel. (C) Generalized erycinine-type boid trunk vertebra in dorsal view. (D) Generalized colubrine-type colubrid trunk vertebra in dorsal view. [A and B modified from Holman, 1981; C, drawing by D. R. Holman; D, drawing by R. Atillio]

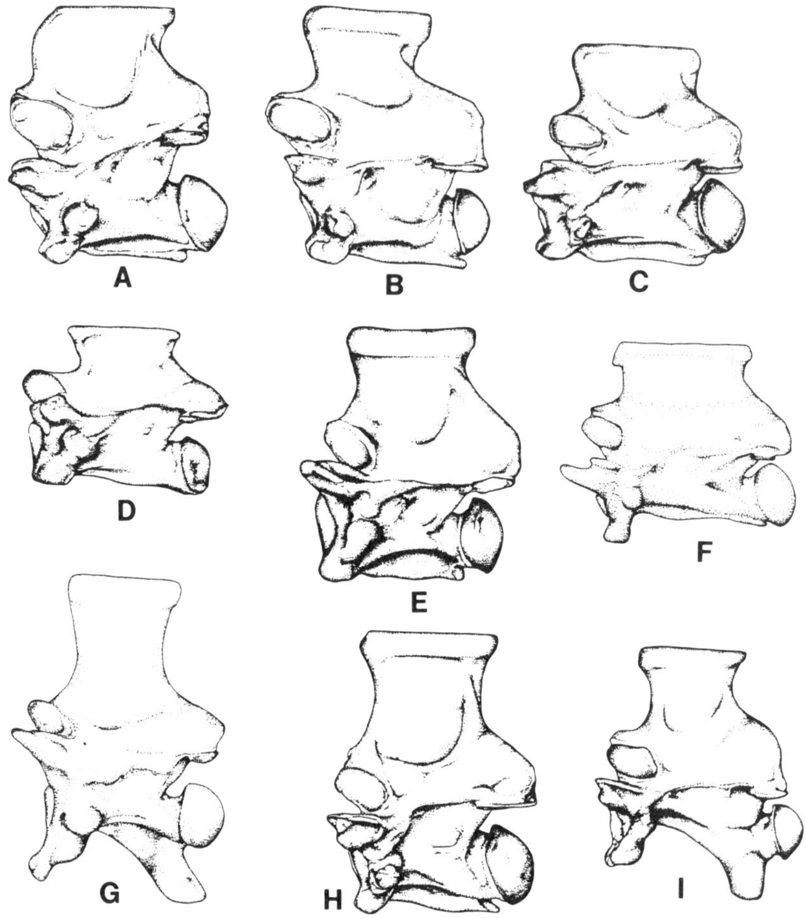

FIGURE 27. Trunk vertebrae of North American colubrid snakes in lateral view. Drawings are from modern skeletons, but all species have been identified from the Pleistocene of North America. (A) *Drymarchon corais*. (B) *Elaphe obsoleta*. (C) *Elaphe vulpina*. (D) *Heterodon platirhinos*. (E) *Lampropeltis getulus*. (F) *Masticophis flagellum*. (G) *Nerodia cyclopion*. (H) *Pituophis melanoleucus mugitus*. (I) *Regina grahamii*. [Drawings by J. Kaminski in Holman, 1981]

squeeze through cracks of a lesser diameter than their own bodies! Specialists can do varying things such as constricting their prey to death, moving very fast, gliding from tree to tree, swimming better than many fishes, climbing, and burrowing exceedingly well.

This is all related to vertebrae that are very structurally complex in their modifications for the functions mentioned and more. The vertebrae that have been utilized most in fossils studies are the trunk vertebrae (Fig. 26–28), defined here as those vertebrae that lie between the small vertebrae of the cervical (neck) region and the structurally different caudal (tail) vertebrae.

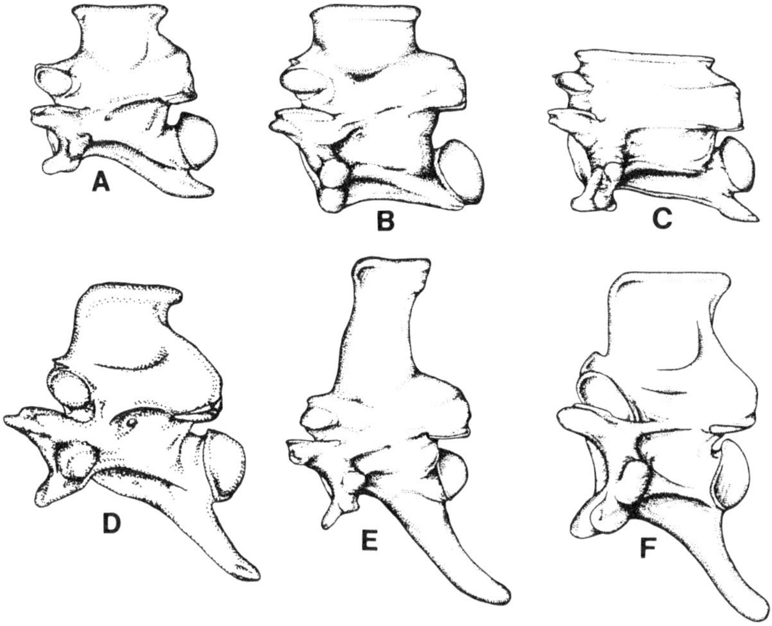

FIGURE 28. Trunk vertebrae of North American colubrid, elapid, and viperid snakes in lateral view. Drawings are from modern skeletons, but these species have been identiifed from the Pleistocene of North America. (A) *Thamnophis butleri* (Colubridae). (B) *Trimorphodon biscutatus* (Colubridae). (C) *Micrurus fulvius* (Elapidae). (D) *Crotalus horridus* (Viperidae). (E) *Crotalus atrox* (Viperidae). (F) *Sistrurus miliarius* (Viperidae) [Drawings by J. Kaminski in Holman, 1981]

Trunk vertebrae in primitive snakes such as the Boidae (Fig. 26C) are usually short and wide with very robust processes, or in advanced snakes such as the Colubridae (Fig. 26D) they are longer and narrower and more slimly built throughout. Very primitive snakes such as the Leptotyphlopidae may have relatively simple vertebrae without neural spines (Fig. 26A), whereas advanced snakes such as the Colubridae and Viperidae may have very high neural spines (e.g., Figs. 27G,H, 28E).

Trunk vertebrae in some snakes bear elongate ventral structures called hypapophyses (Fig. 26B). Elapidae, Viperidae, and some Colubridae have hypapophyses on their trunk vertebrae (e.g., Figs. 27, 28), whereas many other snakes lack hypapophyses on the trunk vertebrae (e.g., most Boidae and Colubridae). A character that is very important in the preliminary identification of snake trunk vertebrae without hypapophyses is whether they have wide or narrow keels on the bottom of the centrum.

When neural spines (Fig. 26A) are present in snakes, very important characters are whether they are high, low, or obsolete, and whether their anterior and posterior margins are straight-sided or indented (see Figs. 27, 28). An equally important character is whether the neural arch has a flattened or vaulted appearance. Finally, the presence or absence of structures on the posterior part of the neural arch called epizygapophyseal spines (Fig. 26A,B) is very important.

Other more subtle characters (Fig. 26A,B), sometimes individually variable, include the shape of the cotyle and the condyle; the presence or absence of pits on either side of the cotyle, and whether they have large or small foramina in them; the shape of the prezygapophyseal and postzygapophyseal faces; the angle that the zygapophyses makes with the long axis of the centrum; the depth of the grooves that lie on either side of the keel on the bottom of the centrum; and the length, thickness, curvature, and terminal shape of the hypapophyses.

There are worrisome variations in the trunk vertebrae of snakes that cause problems in the accurate identification of fossil snakes based on vertebral remains. The problem that many students of fossil snake vertebrae are concerned with (e.g., LaDuke, 1991a) is the intracolumnar variations that occur along the snake vertebral column. Statistical studies have been attempted to rectify this problem from time to time, but these have not been entirely successful (see discussion in LaDuke, 1991a).

These frustrations are compounded by individual vertebral variations within species, as well as the fact that species in many genera have vertebrae that are very similar to one another (e.g., *Thamnophis*, *Storeria*, and *Regina*). In fact, some genera (e.g., *Coluber* and *Masticophis*) are very difficult or impossible to tell apart on the basis of a few vertebral remains. Thus, a comparative collection of modern skeletons with an adequate series of specimens representing each species is probably more necessary for the study of Pleistocene snakes than any other modern herpetological group.

Crocodilians

The only crocodilian remains that have been definitely identified from the Pleistocene of North America have been alligators, and fortunately these large bones are easy to distinguish. Back plates called osteoderms, pieces of skull bones with their characteristic pits, and even vertebrae and limbs may usually be easily identified if a modern alligator skeleton is available.

6

PLEISTOCENE HERPETOFAUNAS OF NORTH AMERICA

One hundred and seventy Pleistocene sites from mainland North America (excluding Mexico) are reviewed in this chapter. To simplify the documentation of these sites, North America has been divided into eight regions (Fig. 29). These regions are based largely on the relative density of herpetofaunal sites in certain areas of North America, and thus they do not always coincide with what would be considered "natural" biotic regions today.

Within each region, fossil herpetofaunal sites are numbered consecutively and designated either as "Sites of Special Interest" or as "Other Important Sites." Several North American sites with less than three herpetological species were omitted from this book if these species were well-documented elsewhere in the region.

Whenever possible, the order of arrangement of the sites in each region is based on the following criteria: (1) age, either absolute or estimated on the basis of fauna or stratigraphy; (2) east to west location of the sites, generally within tiers of states, and (3) north to south location in restricted geographic areas.

Complete faunal lists are provided with each site of special interest, and at the end of each regional account, a summary list of the entire Pleistocene herpetofauna of the region is given with site numbers provided for each identified taxon.

As mentioned in Chapter 2, the Pleistocene is divided into two Land Mammal Ages, the older Irvingtonian Land Mammal Age, which began about 1.9 million years ago, and the younger Rancholabrean Land Mammal Age, which began about about 400,000 years ago. This system was set up on the basis of the biochronology of land mammal faunas. Based on the biochronology of microtine rodents (Repenning, 1987), a system has been devised to further divide the Irvingtonian into two units: Irvingtonian I (IrI), ca. 1.9 to 0.9 million ybp, and Irvingtonian II (IrII), ca. 0.9 to 0.4 million ybp (Repenning, 1987). The Rancholabrean has also been divided into two units (Repenning, 1987): Rancholabrean I (RlbI), ca. 0.4 to 0.15 million ybp, and Rancholabrean II (RlbII), ca. 0.15 million ybp to the present. Many of the Pleistocene herpetolog-

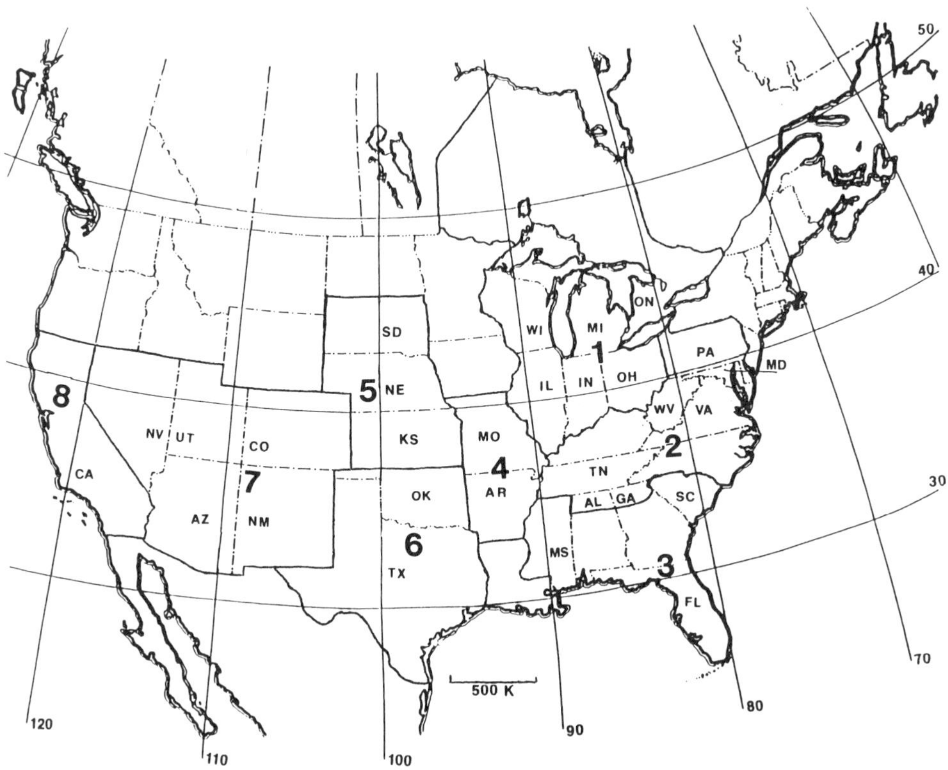

FIGURE 29. North American Pleistocene herpetological regions of this book. The states in the map indicated by United States Postal Service standard abbreviations contain Pleistocene herpetological records reported here. (1) Great Lakes region: Ontario, Michigan, Wisconsin, Ohio, Indiana, Illinois. (2) Appalachian region: Pennsylvania, Maryland, Virginia, West Virginia, Tennessee, Georgia (northern part), Alabama (northern part). (3) Southeastern region: South Carolina, Florida, Georgia (southern part), Alabama (southern part), Mississippi. (4) Ozark region: Missouri, Arkansas. (5) Northern Plains region: South Dakota, Nebraska, Kansas. (6) Southern Plains region: Oklahoma, Texas. (7) Rocky Mountain/Great Basin region: Colorado, New Mexico, Utah, Arizona, Nevada. (8) Pacific Coast region: California.

ical sites that will be discussed here have been assigned to these specific units and others to the broader units, Irvingtonian or Rancholabrean.

In many other cases the faunas have been assigned to specific glacial ages (e.g., Illinoian, Sangamonian, Wisconsinan; see Chapter 2). In the case of Wisconsinan faunas younger than 50,000 ybp, radiocarbon dates are often given.

It is important for the reader to realize that the two preceding systems were set up on the basis of different criteria. North American Land Mammal Ages as used here follow a **biostratigraphy** based on **microtine rodent events** correlated with **potassium-argon-dated ash beds** (Repenning, 1987). Glacial and interglacial ages reflect the stratigraphic relationships of **biotas** to **episodic glacial and interglacial sedimentation**, with absolute radiocarbon dates being applicable beginning at about 50,000 ybp.

Early authors cited in this book mainly used the classic glacial and interglacial age system, whereas modern authors have tended to use Land Mammal Ages. The use of different Pleistocene age terms may be confusing to some nonpaleontological readers. In this book, herpetofaunas designated as **Kansan** may be equated with the **late part of the Irvingtonian Land Mammal Age**, whereas herpetofaunas designated as **Illinoian**, **Sangamonian**, and **Wisconsinan** may be equated respectively with time ranging from the **early to** the **late** part of the **Rancholabrean Land Mammal Age.** Nebraskan, Aftonian, and Yarmouthian herpetofaunas are not currently recognized by most modern authors.

The designated regions of this book are as follow (Fig. 29).

1. Great Lakes: four special and fifteen other sites.
2. Appalachian: twelve special and nine other sites.
3. Southeastern: twelve special and twenty other sites.
4. Ozark: five special and six other sites.
5. Northern Plains: nine special and twelve other sites.
6. Southern Plains: eleven special and fifteen other sites.
7. Rocky Mountain/Great Basin: ten special and sixteen other sites.
8. Pacific Coast: five special and nine other sites.

Great Lakes Region

Because of the glacial history of this region, Pleistocene fossil remains are few, except in two sites. This will be dicussed in more detail later in the book. The Northeastern/Great Lakes region as designated herein includes the following provinces or states that have yielded Pleistocene herpetological material (Fig. 29):

Ontario
Michigan
Wisconsin
Ohio
Indiana
Illinois

Sites of Specials Interest

1. Hopwood Farm Locality, Illinois

The Hopwood Farm locality near Fillmore, Montgomery County, in south-central Illinois is of special interest because three specimens of *Geochelone crassiscuttata* represent the first giant tortoise remains from the Pleistocene of the Great Lakes region and the first published association of this important genus with pollen. The bones accumulated in a kettle in the ridge-drift complex of the

Illinoian till plain where the pollen data documents a vegetational succession from Illinoian glacial to Sangamonian interglacial environments at the type region for each of these stages. The *Geochelone crassiscutata* and other fauna were associated with the Sangamonian interglacial section of the site.

Reference. King and Saunders (1986).

Herpetofauna. *Geochelone crassiscutata*.

2. Sheriden Pit, Ohio

The Sheriden Pit herpetofauna from near Carey, Wyandot County, northwestern Ohio, is of very special interest because it is the only diverse herpetofauna in the Great Lakes drainage basin proper. The cave is located in the Ohio till plain, two miles south of the Defiance moraine. The bones accumulated as a natural cave accumulation by means of the accidental entrapment, as prey items of owls or other predators, or as bones on the surface that were washed into the pit. A carbon-14 date of 11,700 ybp was derived from charcoal at the top of the bone-bearing strata in the cave, thus a late Wisconsinan age is indicated for the upper units of the site, although the lower units may be somewhat older. The fossil herpetofauna consists of fifteen taxa, all of which occur in the area today.

References. Hansen (1992), Holman (1994).

Herpetofauna. *Ambystoma laterale, Bufo americanus, Pseudacris* sp., *Rana clamitans, Rana pipiens, Chelydra serpentina, Chrysemys picta, Emydoidea blandingii, Coluber constrictor, Elaphe vulpina, Lampropeltis triangulum, Opheodrys vernalis, Nerodia sipedon, Regina septemvittata,* and *Thamnophis sirtalis*.

3. Prairie Creek D Fauna, Indiana

The Prairie Creek D fauna of Davies County in southwestern Indiana (Fig. 42) contains a very diverse herpetofauna, also consisting of at least fifteen herpetological species. The Prairie Creek site developed as a low-energy fluviatile feature in a lacustrine drainage system. The Prairie Creek D fauna has been radiocarbon dated at 14,000–15,000 ybp and thus represents the late Wisconsinan. The turtles and snakes of the deposit are of considerable importance from a paleoecological standpoint because all the animals are not only typical of southwestern Indiana today, but are species of common and typical occurrence in southeastern and south-central United States today.

References. Holman (1992a), Holman and Richards (1993).

Herpetofauna. *Ambystoma* sp., *Bufo* sp., *Rana catesbeiana*, *Rana clamitans*, *Rana pipiens* complex, *Rana sylvatica*, *Sternotherus odoratus*, *Chelydra serpentina*, *Chrysemys picta*, *Emydoidea blandingii*, *Graptemys* sp., *Pseudemys* sp., *Apalone spinifera*, *Apalone* sp., *Nerodia erythrogaster*, *Nerodia* sp., and *Thamnophis* sp.

4. Moscow Fissure Site, Wisconsin

The Moscow Fissure site, from near Blanchardville in Iowa County, extreme southwestern Wisconsin, is important because it lies in the Wisconsin driftless area that is believed to have served as a reserve habitat for relict species, some of which dispersed northward after the glacial retreat. The fossil material occurs in sediments between a slump block of dolomite and the parent Ordovician material. The Moscow fauna was deposited near the late Wisconsinan glacial maximum as indicated by a radiocarbon date of 17,050 ± 1,500 ybp. All of the amphibians and reptiles, with the exception of *Bufo americanus* (?) *copei,* a northern subspecies, occur in the area today.

Reference. Foley (1984).

Herpetofauna. *Bufo americanus* (?) *copei*, *Rana* sp., *Elaphe vulpina*, *Lampropeltis triangulum*, and *Thamnophis* sp.

Other Important Sites

5. Kelso Cave, Halton County, Ontario

This site contained a fauna encased in travertine-cemented dolomitic breccia. It is considered of probable Illinoian age based on the presence of the pika, *Ochotona* sp.

Reference. Churcher and Dods (1979).

6. Innerkip Site, Southwestern Ontario

The fauna from this site came from organic silts and peaty deposits beneath two late Wisconsinan tills. Radiocarbon dates provide a minimal middle Wisconsinan age of > 50,000 ybp.

Reference. Pilny and Morgan (1987).

7. Meginity Peccary Cave, Crawford County, Indiana

This cave yielded a fauna from a natural cave accumulation. A radiocarbon date of 30,880 ± 500 ybp is based on peccary bone collagen and indicates a medial Wisconsinan age.

Reference. Richards (1990).

8. Meskill Road Site, St. Clair County, Michigan

This site is based on water-well sediments that were 22.9 m below the surface and below glacial lake clay and silt that was attributed to Glacial Lake Whittlesey (late Wisconsinan).

Reference. Holman, Fisher, and Kapp (1986).

9. Glacial Lake Iroquois Site near Toronto, Ontario

This site is considered to have accumulated in Glacial Lake Iroquois deposits. A radiocarbon date of 11,315 ± 325 ybp was obtained on the antler of an extinct deer, *Torontoceros*, and indicates late Wisconsinan times.

Reference. Churcher and Peterson (1982).

10. Shelton Site, Oakland County, Michigan

The Shelton site accumulated in marginal lake sediments. Radiocarbon dates ranging from 12,320 ± 110 to 11,580 ± 150 ybp were obtained on wood and other organic material, and thus a late Wisconsinan age is indicated.

References. DeFauw and Shoshani (1991), Holman (1992b).

Note. Holman (1992b) questioned the occurrence of *Rana catesbeiana* at this site.

11. New Hudson Site, Oakland County, Michigan

This mastodont site material accumulated in the fluvial sands of a point bar. The site is considered to be late Wisconsinan based on faunal evidence.

Reference. Holman and Fisher (1993).

12. Clark Site, Darke County, Ohio

This fauna accumulated in a kettle bog deposit. Radiocarbon dates indicated an age of 11,040 ± 160 to 10,230 ± 150 ybp and thus a late Wisconsinan designation is in order.

Reference. Holman (1986b).

13. Kolarik Mastodont Site, Starke County, Indiana

This fauna accumulated in a bog deposit. A late Pleistocene age is indicated by the fauna.

Reference. Holman (1992a).

14. Dollens Mastodont Site, Madison County, Indiana

This mastodont site also occurred in a bog setting. It is also believed to be of late Wisconsinan age based on faunal evidence.

Reference. Richards, Whitehead, and Cochran (1987).

15. Christensen Bog Bone Bed, Hancock County, Indiana

This site consisted of a bog burial. Radiocarbon dates of 13,220 ± 100 to 12,060 ± 100 ybp indicate a late Wisconsinan age, but it is believed that the fauna is closer in age to the older date.

Reference. Graham, Holman, and Parmalee (1983).

16. Anderson Pit Cave, Monroe County, Indiana

This fauna is the result of a natural cave accumulation. Based on the association of the fauna with the extinct armadillo *Dasypus bellus*, it is believed that the cave fauna was a late Wisconsinan assemblage.

References. Holman and Richards (1981), Richards (1990).

17. Mill Cave, Green County, Indiana

This fauna is also the result of a natural cave accumulation. It is believed to be late Wisconsinan based on the faunal content.

Reference. Richards (1984).

18. Polecat Creek Local Biota, Coles County, Illinois

It is assumed that this was a fluvial deposit representing the late Wisconsinan.

References. Galbreath (1938), Preston (1979).

19. Clear Creek Deposit, Sangamon County, Illinois

These remains accumulated in fluvial gravels. The site possibly represents part of the Mackinaw Member of the Henry Formation. It is considered to be of late Wisconsinan age based on faunal remains.

References. Holman (1966a), Holman (1992a).

Summary of Herpetological Taxa*

Ambystomatidae
- *Ambystoma laterale:* 2
- *Ambystoma* sp.: 3

Bufonidae
- *Bufo americanus* (?) *copei:* 4
- *Bufo americanus:* 2, 8
- *Bufo* sp. (?) *Bufo americanus:* 5
- *Bufo* sp.: 3

Hylidae
- *Pseudacris* sp.: 2

Ranidae
- *Rana catesbeiana:* 3, 10 (see comments for site 10)
- *Rana clamitans:* 2, 3, 10
- *Rana pipiens:* 2
- *Rana pipiens* complex: 3, 13, 15
- *Rana sylvatica:* 3
- *Rana* sp.: 4, 9, 14

Kinosternidae
- *Sternotherus odoratus:* 3

Chelydridae
- *Chelydra serpentina:* 2, 3, 12, 14, 15, 19

Emydidae
- *Chrysemys picta:* 2, 3, 11, 12, 15, 18
- *Emydoidea blandingii:* 2, 3, 6
- *Graptemys* sp.: 3
- *Pseudemys* sp.: 3
- *Terrapene carolina:* 18
- *Trachemys scripta:* 18, 19

Testudinidae
- *Geochelone crassiscutata:* 1

*Numerals in summary correspond to sites described in this section.

Trionychidae
Apalone spinifera: 3, 19
Apalone sp.: 3, 15
Anguidae
Ophisaurus cf. *Ophisaurus attenuatus:* 17
Colubridae
Coluber constrictor: 2
Elaphe vulpina: 2, 4
Lampropeltis triangulum: 2, 4
Opheodrys vernalis: 2, 16
Nerodia erythrogaster: 3
Nerodia sipedon: 2
Nerodia sp.: 3
Regina septemvittata: 2
Thamnophis sirtalis: 2
Thamnophis sp.: 3, 4
Viperidae
Crotalus horridus: 7, 16

Appalachian Region

The Appalachian region contains an abundance of caves and related structures, and for that reason has yielded a relatively large number of diverse herpetofaunas. The following states that have appalachian topography within their borders are included in this section (Fig. 29):

Pennsylvania
Maryland
Virginia
West Virginia
Tennessee
Georgia (northern part)
Alabama (northern part)

Sites of Special Interest

1. Cumberland Cave, Maryland

The Cumberland Cave site in Allegany County, Maryland, has provided a large window on the pre-Wisconsinan herpetological life in North America east of the Mississippi and north of Florida. The vertebrate assemblage represents a natural cave filling. The site represents the Irvingtonian II Land Mammal Age. Repenning (1987) and Holman (1977a) referred to the site as Kansan.

At least thirty species of salamanders, anurans, chelonians, lizards, and snakes are represented. The modern nature of the herpetofauna is striking, as

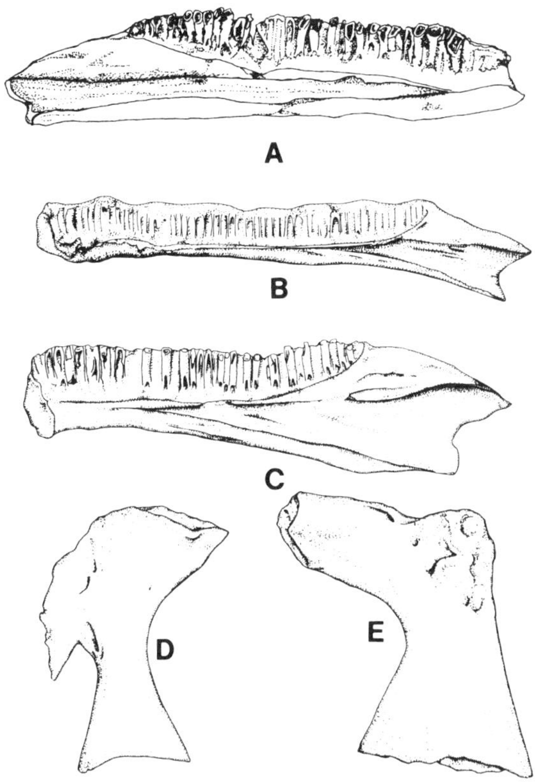

FIGURE 30. *Cryptobranchus guildayi* Holman. (A) Holotype left dentary (Carnegie Museum of Natural History Number 20470) from the Irvingtonian of Cumberland Cave, Allegany County, Maryland [from Holman, 1977a]. (B, C) Right dentaries; (E) scapula, and (G) epihyal of *Cryptobranchus guildayi* from the Irvingtonian of Trout Cave, Pendleton County, West Virginia [from Holman, 1982]. Vertebrae of this species from New Trout Cave are shown in Figure 16A.

only one taxon, *Cryptobranchus guildayi* Holman, 1977a, which was described as new in this fauna, represents an extinct form. This species (Fig. 30) was described on the basis of a left dentary that differed from the modern species *Cryptobranchus alleganiensis* in having a much longer Meckelian groove and in having the dentary more weakly curved than in the modern species. This questionable species will be discussed in Chapter 8.) *Ambystoma tigrinum* is a locally absent species in the area today. *Elaphe vulpina*, is a western extralimital form. The only North American fossil record of the presently endangered species *Clemmys muhlenbergii* occurs in this fauna.

References. Brattstrom (1954b), Lynch (1966), Holman (1977a).

Herpetofauna. *Cryptobranchus guildayi* (**type locality**) *Notophthalmus* cf. *Notophthalmus viridescens, Ambystoma maculatum, Ambystoma opacum, Ambystoma tigrinum, Gyrinophilus* cf. *Gyrinophilus porphyriticis, Plethodon* *glutinosus, Pseudotriton ruber, Bufo americanus, Bufo woodhousii fowleri, Pseudacris crucifer, Pseudacris triseriata, Rana clamitans, Rana pipiens* complex, *Rana sylvatica, Chrysemys picta, Clemmys muhlenbergii, Terrapene carolina, Sceloporus* cf. *Sceloporus undulatus, Eumeces* cf. *Eumeces fasciatus, Carphophis amoenus, Coluber constrictor, Diadophis punctatus, Elaphe vulpina, Heterodon* cf. *Heterodon platirhinos, Lampropeltis triangulum, Nerodia sipedon, Opheodrys vernalis, Storeria* sp., *Thamnophis* sp., and *Crotalus horridus.*

2. Hamilton Cave, West Virginia

The Hamilton Cave herpetofauna lies in Pendleton County in eastern West Virginia. This herpetofauna is the most taxonomically diverse Irvingtonian herpetofauna known. Many of the amphibian and reptile bones may have been derived from coprolites as the bones were fragmentary compared to those of the large mammalian species recovered from the cave. The fauna is believed to represent the Irvingtonian II Land Mammal Age of Repenning (1987). None of the herpetological taxa are extinct, and all of them could be found living in or near Pendleton County, West Virginia, today, with the possible exception of the *Gyrinophilus* fossils which might represent a neotenic cave form such as *Gyrinophilus palleucus* (see Fig. 4E,F) and *Cryptobranchus* sp., which is locally absent.

Reference. Holman and Grady (1989).

Herpetofauna. *Cryptobranchus* sp., *Notophthalmus viridescens, Ambystoma jeffersonianum* complex, *Ambystoma maculatum, Ambystoma opacum, Ambystoma tigrinum, Desmognathus fuscus, Desmognathus monticola, Desmognathus ochrophaeus, Gyrinophilus porphyriticus, Gyrinophilus* sp. (?cave neotene), *Bufo americanus, Bufo woodhousii fowleri, Hyla chrysoscelis* or *versicolor, Pseudacris crucifer, Rana catesbeiana, Rana clamitans, Rana pipiens* complex *Rana sylvatica,* Emydidae indet., *Sceloporus undulatus, Eumeces fasciatus, Eumeces laticeps, Carphophis amoenus, Coluber constrictor, Elaphe* sp., *Lampropeltis triangulum, Nerodia sipedon, Storeria* sp., *Thamnophis* sp., and *Crotalus horridus.*

3. New Trout Cave, West Virginia

The New Trout Cave site, in Pendleton County in eastern West Virginia (not to be confused with Trout Cave [site 16 in this account] in the same county in West Virginia) is of special significance because it is stratified, and each of the strata has been radiocarbon dated. The herpetological bones in the cave are the result of a natural cave accumulation. The strata occurred in intervals between the closing of a natural opening in the roof of the cave.

Unit I has a radiocarbon date of 17,060 $\pm$ 220 ybp. Unit II has radiocarbon dates of 28,250 $\pm$ 850 and 29,400 $\pm$ 1,700 ybp. Unit III represents levels below Units I and II and represents a time greater than 30,000 ybp. None of the herpetologic species are extinct in any of the units, whereas the mammalian fauna has several extinct species at all levels. Moreover, herpetologic species that do not occur in the areas today are only locally absent. Twenty-seven of the twenty-eight amphibian and reptile species in Units I and II occur in the cave area today. Seventeen of the twenty-one herpetological taxa in Unit III occur in the area today.

Reference. Holman and Grady (1987).

Herpetofauna. Unit I: *Notophthalmus viridescens, Ambystoma jeffersonianum* complex, *Ambystoma maculatum, Ambystoma opacum, Desmognathus fuscus, Desmognathus monticola, Desmognathus ochrophaeus, Gyrinophilus porphyriticus, Plethodon* cf. *Plethodon glutinosus, Pseudotriton* cf. *Pseudotriton ruber, Bufo* sp., *Pseudacris crucifer, Rana clamitans, Rana pipiens* complex, *Rana sylvatica, Sceloporus undulatus, Eumeces laticeps, Carphophis amoenus, Coluber constrictor, Diadophis punctatus, Elaphe* sp., *Heterodon platirhinos, Lampropeltis triangulum, Nerodia sipedon, Storeria* sp., *Thamnophis brachystoma, Thamnophis sirtalis, Thamnophis* sp., and *Crotalus horridus.* Unit II: *Notophthalmus viridescens, Ambystoma jeffersonianum* complex, *Ambystoma maculatum, Ambystoma opacum, Desmognathus fuscus, Desmognathus monticola, Desmognathus ochrophaeus, Gyrinophilus porphyriticus, Pseudotriton* cf. *Pseudotriton ruber, Bufo* sp., *Rana pipiens* complex, *Rana sylvatica, Carphophis amoenus, Diadophis punctatus, Elaphe* sp., *Heterodon platirhinos, Lampropeltis triangulum, Nerodia sipedon, Storeria* sp., *Thamnophis brachystoma, Thamnophis sirtalis,* and *Thamnophis* sp. Unit III: *Cryptobranchus* sp., *Ambystoma jeffersonianum* complex, *Ambystoma maculatum, Desmognathus fuscus, Desmognathus ochrophaeus, Bufo* sp., *Rana pipiens* complex, *Emydoidea blandingii, Sceloporus undulatus, Carphophis amoenus, Coluber constrictor, Elaphe vulpina, Elaphe* sp., *Heterodon platirhinos, Lampropeltis triangulum, Storeria* sp., *Thamnophis brachystoma, Thamnophis sirtalis, Thamnophis* sp., *Agkistrodon contortrix,* and *Crotalus horridus.*

4. Strait Canyon Fissure, Virginia

The Strait Canyon Fissure site lies in Highland County, in the Appalachians of Virginia. It is from one of the earlier Wisconsinan vertebrate faunas in the Appalachians. The site developed in a fissure fill. Radiocarbon dating yielded an age of 29,870 + 1,800 to 29,870 − 1,400 ybp. None of these amphibians and reptiles represent extinct forms. Fay (1984) reports that although all of these species do not occur in the area today, they all coexist 150 km to the east in the Piedmont.

Reference. Fay (1984).

Herpetofauna. *Ambystoma maculatum, Desmognathus* (?) *fuscus, Plethodon glutinosus, Bufo americanus, Bufo woodhousii fowleri, Rana catesbeiana, Rana clamitans, Rana pipiens* complex, *Rana sylvatica, Terrapene carolina, Trachemys* or *Pseudemys,* (?) *Eumeces* sp., *Coluber* or *Masticophis, Diadophis punctatus, Elaphe guttata, Elaphe obsoleta, Lampropeltis calligaster, Lampropeltis getula, Lampropeltis triangulum, Nerodia sipedon, Opheodrys vernalis, Storeria* (?) *occipitomaculata, Thamnophis sirtalis, Agkistrodon contortrix,* and *Crotalus horridus.*

5. Bell Cave, Alabama

The Bell Cave fauna in Colbert County in northwestern Alabama represents one of the few stratified cave sites where Pleistocene amphibians and reptiles have been recovered. The herpetological fossils were derived as a natural cave accumulation. Three Wisconsinan units have been identified. Unit I has a radiocarbon date of 11,820 + 480 to 11,820 − 500 ybp. Unit II, below Unit I, did not have sufficient bone for a radiocarbon sample. Unit III has a radiocarbon date of 26,500 + 870 to 26,500 − 990 ybp.

The wide variety of habitats represented by the fossils is attributed to transportation by paleopredators. Excessive damage was present in many of the fossils due to predators and gnawing by scavengers. None of the amphibians or reptile species is extinct in contrast to the mammalian faunas, which have several extinct taxa. Only a very few taxa that occur in the Bell Cave herpetofauna have not been recorded in the immediate vicinity of the cave today. *Elaphe vulpina* is a rather striking western extralimital form.

Reference. Holman, Bell, and Lamb (1990).

Herpetofauna. Unit I: *Cryptobranchus alleganiensis, Ambystoma* sp., *Ambystoma maculatum, Eurycea* sp., *Plethodon glutinosus, Bufo americanus, Bufo woodhousii fowleri, Bufo* sp., *Hyla gratiosa, Rana pipiens* complex, *Graptemys geographica, Pseudemys concinna, Trachemys* or *Pseudemys, Apalone spinifera, Apalone* sp., *Carphophis amoenus, Elaphe* cf. *Elaphe guttata, Elaphe vulpina, Elaphe* sp., *Heterodon* cf. *Heterodon platirhinos, Lampropeltis getula, Lampropeltis triangulum triangulum, Nerodia sipedon, Nerodia* sp., *Opheodrys aestivus, Storeria* sp., *Thamnophis sirtalis, Thamnophis* sp., and *Crotalus horridus*. Unit II: *Cryptobranchus alleganiensis, Ambystoma* sp., *Ambystoma maculatum, Plethodon glutinosus, Bufo americanus, Bufo woodhousii fowleri, Bufo* sp., *Rana catesbeiana, Rana pipiens* complex, *Graptemys geographica, Pseudemys concinna, Apalone* sp., *Carphophis amoenus, Coluber constrictor, Diadophis punctatus, Elaphe vulpina, Elaphe* sp., *Heterodon* cf. *Heterodon platirhinos, Lampropeltis getula, Lampropeltis triangulum triangulum, Nerodia sipedon, Nerodia* sp., *Opheodrys aestivus, Storeria* sp., *Thamnophis sirtalis, Thamnophis* sp., and *Crotalus horridus*. Unit III: *Cryptobranchus alleganiensis, Desmognathus ochrophaeus, Desmognathus* sp., *Eurycea* sp., *Rana pipiens* complex, *Rana* sp., *Chrysemys picta, Anolis carolinensis, Heterodon* cf. *Heterodon platirhinos, Lampropeltis getula, Lampropeltis triangulum triangulum, Nerodia sipedon, Thamnophis sirtalis, Thamnophis* sp., *Agkistrodon* sp., and *Crotalus horridus*.

6. Guy Wilson Cave, Tennessee

The Guy Wilson Cave is in Sullivan County in extreme eastern Tennessee. It is the richest Pleistocene herpetofauna known from this particular part of the Appalachian region. This cave formed a natural trap for vertebrate animals. A radiocarbon date of 19,700 ± 600 ybp is associated with the fauna and thus a late Wisconsinan age is represented. Again, the herpetofauna is very similar to one that would be found in the area today.

Reference. This list was provided through the kindness of Dr. Leslie P. Fay, who has studied the fauna. I have modified his taxonomic designations somewhat to conform to the format of this book.

Herpetofauna. *Cryptobranchus alleganiensis*, *Necturus maculosus*, *Ambystoma maculatum* complex, *Ambystoma* sp., *Desmognathus* or *Leurognathus*, *Gyrinophilus* or *Pseudotriton*, *Gyrinophilus porphyriticus*, *Plethodon* sp., complex, *Pseudotriton* sp., *Bufo americanus*, *Bufo woodhousii fowleri*, *Bufo* sp., *Hyla* sp., *Rana pipiens* complex, *Rana* sp., (?) *Chrysemys picta*, (?) *Terrapene carolina*, *Sceloporus undulatus*, *Eumeces* cf. *Eumeces laticeps*, *Coluber* or *Masticophis*, *Elaphe guttata*, *Elaphe obsoleta*, (?) *Heterodon platirhinos*, *Lampropeltis getula*, *Lampropeltis triangulum*, *Nerodia sipedon*, *Thamnophis sirtalis*, (?) *Virginia valeriae*, and *Crotalus horridus*.

7. Cheek Bend Cave, Tennessee

The Cheek Bend Cave, in Maury County in the Nashville Basin of southern central Tennessee, provides the largest Pleistocene herpetofauna known from this particular part of the Appalachian region. The fossil bones represent a natural cave accumulation, at least in the Pleistocene strata. Based on radiocarbon dates the Pleistocene herpetofauna accumulated between 16,000 and 12,000 ybp in the late Wisconsinan. This is a multifaceted site with all classes of vertebrates represented as well as artifacts and Holocene as well as Pleistocene strata.

In general the Pleistocene herpetofauna represents animals that could be found in or at least near Maury County, Tennessee today. Nevertheless, southern extralimital forms *Bufo* cf. *Bufo terrestris* and *Rhadinaea flavilata* occurred with the northern extralimital form *Clemmys insculpta*, all in the same stratum (Unit III). The faunal list in this book comes from two separate works. Salamanders are from Miller (1992) and anurans, turtles, lizards, and snakes from Klippel and Parmalee (1982).

References. Klippel and Parmalee (1982), Miller (1992).

Herpetofauna. *Cryptobranchus alleganiensis, Ambystoma maculatum, Ambystoma tigrinum, Ambystoma* sp., *Eurycea* cf. *Eurycea bislineata*, *Eurycea lucifuga*, *Eurycea* sp., *Bufo* cf. *Bufo americanus*, *Bufo* cf. *Bufo terrestris*, *Bufo* cf. *Bufo woodhousii*, *Bufo* sp., *Pseudacris crucifer*, *Rana* (several species), *Clemmys insculpta*, *Eumeces fasciatus*, *Carphophis amoenus*, *Diadophis punctatus*, *Lampropeltis getula*, *Opheodrys* sp., *Rhadinaea flavilata*, *Storeria dekayi*, *Storeria* sp., *Thamnophis sirtalis*, and *Virginia* sp.

8. New Paris 4 Site, Pennsylvania

The New Paris 4 site of Bedford County in southern central Pennsylvania is one of the classic herpetological sites in North America. It has been studied by Guilday, Martin, and McCrady (1964), Lynch (1966), and most recently by Fay (1988), who summarized the earlier lists and added many species to them. The fossils were deposited in a cave fissure fill by natural means. A radiocarbon date of 11,300 ± 1,000 ybp was taken from Unit I above the boundary of Unit II at this site and thus a late Wisconsinan age is indicated. All the herpetological species identified from this cave represent living species, but several of them occur farther south in the Appalachian region today (e.g., *Lampropeltis calligaster, Lampropeltis getula, Opheodrys aestivus, Pituophis melanoleucus*).

References. Guilday, Martin, and McCrady (1964), Lynch (1966), Fay (1988).

Herpetofauna. Unit I: *Notophthalmus viridescens*, *Ambystoma maculatum* complex, *Ambystoma* sp., *Desmognathus* or *Leurognathus*, *Eurycea* sp., *Plethodon* sp. complex, *Bufo americanus*, *Pseudacris crucifer*, *Rana* (?) *clamitans*, *Rana* (?) *palustris*, *Rana pipiens* complex, *Rana sylvatica*, *Carphophis amoenus*, *Coluber* or *Masticophis*, *Diadophis punctatus*, *Elaphe guttata*, *Elaphe obsoleta*, *Lampropeltis calligaster*, *Lampropeltis getula*, *Lampropeltis triangulum*, *Nerodia sipedon*, *Opheodrys aestivus*, *Pituophis melanoleucus*, (?) *Regina septemvittata*, *Storeria dekayi*, *Storeria occipitomaculata*, *Thamnophis* (?) *sauritus*, *Thamnophis sirtalis*, *Thamnophis* sp., *Virginia valeriae*, *Agkistrodon contortrix*, and *Crotalus horridus*. Unit II: *Notophthalmus viridescens*, *Ambystoma maculatum* complex, *Ambystoma* sp., *Desmognathus* or *Leurognathus*, *Eurycea* sp., *Plethodon* sp. complex, *Pseudotriton* sp., *Bufo americanus*, *Bufo americanus* cf. *Bufo americanus copei*, *Bufo woodhousii fowleri*, *Pseudacris crucifer*, *Rana* (?) *palustris*, *Rana pipiens* complex, *Rana sylvatica*, *Diadophis punctatus*, *Elaphe obsoleta*, *Lampropeltis getula*, *Nerodia sipedon*, *Pituophis melanoleucus*, *Storeria dekayi*, *Thamnophis sirtalis*, *Thamnophis* sp., *Virginia valeriae*, *Agkistrodon contortrix*, and *Crotalus horridus*.

9. Clark's Cave, Virginia

Clark's Cave, in Bath County in Northwestern Virginia, is one of two very large herpetofaunas in the Virginia Appalachians. The site of the deposition of the bones was an owl roost. A radiocarbon date of ca. 11,000 ybp indicates a very late Wisconsinan age for this deposit. All herpetological species of the Clark's Cave site are extant today and only five are extralimital (e.g., the northern *Opheodrys vernalis* and southern *Virginia striatula*).

References. Fay (1988)

Herpetofauna. *Notophthalmus viridescens*, *Ambystoma maculatum* complex, *Ambystoma opacum* complex, *Ambystoma* sp., *Desmognathus* or *Leurognathus*, *Eurycea* sp., *Gyrinophilus porphyriticus*, *Plethodon* complex, *Pseudotriton* sp., *Scaphiopus holbrookii*, *Bufo americanus*, *Bufo woodhousii fowleri*, *Pseudacris crucifer*, *Rana catesbeiana*, *Rana clamitans*, *Rana palustris*, *Rana pipiens* complex, *Rana sylvatica*, *Graptemys geographica*, *Sceloporus undulatus*, *Eumeces* (?) *laticeps*, *Carphophis amoenus*, *Cemophora coccinea*, *Coluber* or *Masticophis*, *Diadophis punctatus*, *Elaphe guttata*, *Elaphe obsoleta*, *Heterodon platirhinos*, *Lampropeltis calligaster*, *Lampropeltis getula*, *Lampropeltis triangulum*, *Nerodia sipedon*, *Opheodrys aestivus*, *Opheodrys vernalis*, (?) *Regina septemvittata*, *Storeria dekayi*, *Storeria occipitomaculata*, *Thamnophis* (?) *sauritus*, *Thamnophis sirtalis*, *Virginia striatula*, *Virginia valeriae*, and *Crotalus horridus*.

10. Natural Chimneys Site, Virginia

The Natural Chimneys site in Augusta County, northwestern Virginia, is the second of the very large Pleistocene herpetofaunas in the Virginia Appalachians. The herpetofauna from this site was also derived by means of an owl roost. The site is considered to be of the very latest Wisconsinan age on the basis of the vertebrate fauna. All the amphibians and reptiles from the Natural Chimneys site represent extant taxa. Five herpetological taxa do not occur in the area of the Natural Chimneys site today, but they reach the latitude of this site in other areas of the Carolinean biotic province. Fay (1988) eliminated some of the herpetological records recorded by Guilday (1962) and added many new records.

References. Guilday (1962), Fay (1988).

Herpetofauna. *Notophthalmus viridescens*, *Ambystoma maculatum* complex, *Ambystoma opacum* complex, *Ambystoma tigrinum*, *Desmognathus* or *Leurognathus*, *Eurycea* sp., *Gyrinophilus porphyriticus*, *Plethodon* sp. complex, *Pseudotriton* sp., *Scaphiopus holbrookii*, *Bufo americanus*, *Bufo woodhousii fowleri*, *Rana catesbeiana*, *Rana clamitans*, *Rana* (?) *palustris*, *Rana pipiens* complex, *Rana sylvatica*, *Chelydra serpentina*, *Terrapene carolina*, *Sceloporus undulatus*, *Scincella lateralis*, *Carphophis amoenus*, *Cemophora coccinea*, *Coluber* or *Masticophis*, *Diadophis punctatus*, *Elaphe guttata*, *Elaphe obsoleta*, *Lampropeltis getula*, *Lampropeltis triangulum*, *Nerodia* (?) *erythrogaster*, *Nerodia sipedon*, *Opheodrys vernalis*, (?) *Regina septemvittata*, *Storeria dekayi*, *Storeria occipitomaculata*, *Thamnophis sirtalis*, *Thamnophis* sp., *Virginia striatula*, *Virginia valeriae*, *Agkistrodon contortrix*, and *Crotalus horridus*.

11. Kingston Saltpeter Cave, Georgia

The Kingston Saltpeter Cave site in Bartow County, in northwestern Georgia, is one of the two very large Pleistocene herpetofaunas of the Georgia Appalachians. The amphibian and reptile bones are thought to have accumulated principally through the activity of raptors. A radiocarbon date of 10,300 $\pm$ 130 ybp on bone collagen places this site very late in the Wisconsinan. All the herpetological species of this very rich site are extant and the fauna is typical of one that could be found in the area today.

Reference. Fay (1988).

Herpetofauna. *Notophthalmus viridescens*, *Ambystoma maculatum* complex, *Ambystoma opacum* complex, *Ambystoma tigrinum*, *Desmognathus* or *Leurognathus*, *Eurycea* sp., *Gyrinophilus* sp., *Plethodon* sp. complex, *Pseudotriton ruber*, *Scaphiopus holbrookii*, *Bufo americanus*, *Bufo woodhousii fowleri*, *Pseudacris crucifer*, *Gastrophryne carolinensis*, *Rana catesbeiana*, *Rana clamitans*, *Rana pipiens* complex, *Rana* sp., Kinosternidae indet., *Chrysemys picta*, *Terrapene carolina*, *Trachemys* or *Pseudemys*, large Emydidae indet., *Anolis carolinensis*, *Sceloporus undulatus*, *Scincella lateralis*, *Carphophis amoenus*, *Coluber* or *Masticophis*, *Diadophis punctatus*, *Elaphe guttata*, *Elaphe obsoleta*, *Lampropeltis calligaster*, *Lampropeltis getula*, *Lampropeltis triangulum*, *Nerodia sipedon*, *Opheodrys* (?) *aestivus*, *Pituophis melanoleucus*, *Storeria dekayi*, (?) *Storeria* sp., *Thamnophis* (?) *sauritus*, *Thamnophis sirtalis*, *Virginia striatula*, *Virginia valeriae*, and *Agkistrodon contortrix*.

12. Ladds Quarry Site, Georgia

The Ladds Quarry site in Bartow County in northwestern Georgia is the second very large herpetofauna in the Georgia Appalachians. It consists of a system of caves and fissure fills, some of which may not be contemporaneous. Various agents, including jaguars, are responsible for the accumulation of the bones at this imnportant site. Radiocarbon dates on turtle bone collagen from two fissure fills, called the "Tapir site" and the "Jaguar site," were 10,290 ± 100 ybp (Tapir) and 10,940 ± 210 ybp (Jaguar) and thus a very late Wisconsinan age is indicated. The Ladds herpetofauna, including both the dated sites, contained extralimital forms from both the north (*Clemmys insculpta*, *Elaphe vulpina*, *Lampropeltis triangulum*) and the south (*Bufo terrestris*, *Pseudemys nelsoni*, and the extinct giant land tortoise, *Geochelone crassiscutata*), contrasting with the nearby Kingston Saltpeter Cave. Some of the fossil turtle shells had damage typical of jaguar predation.

References. Wilson (1975), Holman (1967, 1985a,b).

Herpetofauna. *Ambystoma tigrinum*, *Desmognathus* or *Leurognathus*, *Gyrinophilus* sp., *Plethodon glutinosus*, *Pseudotriton ruber*, *Bufo americanus*, *Bufo terrestris*, *Hyla chrysoscelis* or *versicolor*, *Pseudacris crucifer*, *Gastrophryne carolinensis*, *Rana utricularia*, *Kinosternon subrubrum*, *Chrysemys picta*, *Clemmys insculpta*, *Pseudemys concinna*, *Pseudemys nelsoni*, *Terrapene carolina carolina*, *Geochelone crassiscutata*, *Anolis carolinensis*, *Sceloporus undulatus*, *Carphophis amoenus*, (?) *Coluber constrictor*, *Diadophis punctatus*, *Elaphe vulpina*, *Heterodon platirhinos*, *Lampropeltis calligaster*, *Lampropeltis getula*, *Lampropeltis triangulum triangulum*, *Nerodia sipedon*, *Storeria dekayi*, *Thamnophis sirtalis*, *Agkistrodon piscivorous*, and *Crotalus horridus*.

Other Important Sites

13. Hanover Quarry Number 1, Fissure, Adams County, Pennsylvania

This herpetofauna accumulated in a fissure fill. It is believed to represent the Irvingtonian Land Mammal Age, possibly the early part.

Reference. A list of species from this site was kindly provided by Dr. Leslie P. Fay.

14. Port Kennedy Sinkhole, Montgomery County, Pennsylvania

This vertebrate fauna accumulated as a natural sinkhole filling. The site represents the Irvingtonian Land Mammal Age (Woodburne, 1987). An important fossil is "cf. *Geochelone* sp.," an extinct land tortoise.

Reference. Daeschler, Spamer, and Parris (1993).

15. Trout Cave, Pendleton County, West Virginia

The Trout Cave fauna (not to be confused with the younger New Trout Cave [site 6 in this section] of the same county in West Virginia) is a natural cave accumulation. It is considered to represent the Irvingtonian Land Mammal based on the mammalian fauna, and Holman (1982) refers to it as a Kansan deposit. This site is important in that it contains additional material of the controversial extinct salamander *Cryptobranchus guildayi* (Figs. 16A, 30).

Reference. Holman (1982).

16. Frankstown Cave, Blair County, Pennsylvania

The Frankstown Cave fauna accumulated as a fissure fill. It is considered to be of Wisconsinan age. Fay (1988) recently revised the herpetofaunal list of this classic site.

References. Richmond (1964), Fay (1988).

17. Bootlegger Sink, York County, Pennsylvania

This herpetological assemblage occurred naturally in a bone breccia in a sinkhole. It has been designated as a very late Wisconsinan assemblage. I have modified some of the taxonomic designations (list at the end of this account) proposed by Neil D. Richmond in Guilday, Hamilton, and McCrady (1966).

Reference. Guilday et al. (1966).

18. Saltville Site, Saltville Valley, Virginia

This site developed in fluvial and lentic sediments in the late Wisconsinan. The herpetofauna comes from four radiocarbon dated units: I, 27,000 to 14,500 ybp; II, ca. 14,500 to 13,500 ybp; III, 13,500 to 10,500 ybp; IV, 10,500 to 10,000 ybp.

Reference. Holman and McDonald (1986).

19. Baker Bluff Cave, Sullivan County, Tennessee

This cave site produced a fauna that was the result of predation by owls and the activities of wood rats (*Neotoma*). The herpetofauna listed here (summary at end of section) comes from the mid-levels of the site, which date from 11,640 to 10,560 ybp.

References. Van Dam (1978), Fay (1988).

20. Robinson Cave, Overton County, Tennessee

This herpetofauna is the result of a natural cave accumulation. It has been judged to represent the late Wisconsinan based on the composition of the fauna.

References. Lynch (1966), Guilday, Hamilton, and McCrady (1969).

21. First American Bank Site, Davidson County, Tennessee

This site developed as a natural cave accumulation. The small list of reptile fossils included here in the regional summary comes from a zone where large, extinct mammals were found and that was radiocarbon dated as from 10,035 $\pm$ 650 to 9,410 $\pm$ 155 ybp. Thus the bones represent very latest Wisconsinan or very earliest Holocene times.

Reference. Guilday (1977).

Summary of Herpetological Taxa*

Cryptobranchidae
Cryptobranchus alleganiensis: 5-I, 5-II, 5-III, 6, 7, 18-III, 19
Cryptobranchus guildayi: 1, 15
Cryptobranchus sp.: 2, 3-III
Salamandridae
Notophthalmus viridescens: 2, 3-I, 3-II, 8-I, 8-II, 9, 10, 11, 16.
Notopthalmus cf. *viridescens:* 1, 18-II
Proteidae
Necturus maculosus: 6, 19
Ambystomatidae
Ambystoma jeffersonianum complex: 2, 3-I, 3-II, 3-III
Ambystoma maculatum: 1, 2, 3-I, 3-II, 3-III, 4, 5-I, 5-II, 7
Ambystoma maculatum complex: 6, 8-I, 8-II, 9, 10, 11, 19
Ambystoma opacum: 1, 2, 3-I, 3-II, 15
Ambystoma opacum complex: 9, 10, 11, 19
Ambystoma tigrinum: 1, 2, 7, 10, 11, 12
Ambystoma sp.: 5-I, 5-II, 6, 7, 8-I, 8-II, 9, 16, 19, 20
Plethodontidae
Desmognathus fuscus: 2, 3-I, 3-II, 3-III
Desmognathus (?) *fuscus:* 4
Desmognathus monticola: 2, 3-I, 3-II
Desmognathus ochrophaeus: 2, 3-I, 3-II, 3-III, 5-III
Desmognathus sp.: 5-III
Desmognathus or *Leurognathus:* 6, 8-I, 8-II, 9, 10, 11, 12, 19
Eurycea cf. *Eurycea bislineata:* 7
Eurycea lucifuga: 7
Eurycea sp.: 5-I, 5-III, 7, 8-I, 8-II, 9, 10, 11
Gyrinophilus porphyriticus: 2, 3-I, 3-II, 6, 9, 10
Gyrinophilus cf. *Gyrinophilus porphyriticus:* 1
Gyrinophilus sp.: 2, 11, 12
Gyrinophilus or *Pseudotriton:* 6
Plethodon glutinosus: 1, 4, 5-I, 5-II, 12
Plethodon cf. *Plethodon glutinosus:* 3-I
Plethodon sp. complex: 6, 8-I, 8-II, 9, 10, 11
Pseudotriton ruber: 1, 11, 12
Pseudotriton cf. *Pseudotriton ruber:* 3-I, 3-II
Pseudotriton sp.: 6, 8-II, 9, 10
Caudata indet. 17
Pelobatidae
Scaphiopus holbrookii: 9, 10, 11, 17

---

*Numerals in summary correspond to sites described in this section.

Scaphiopus sp.: 20
Bufonidae
Bufo americanus cf. *Bufo americanus copei:* 8-II
Bufo americanus: 1, 2, 4, 5-I, 5-II, 6, 8-I, 8-II, 9, 10, 11, 12, 13, 15, 16, 17, 19, 20
Bufo cf. *Bufo americanus:* 7, 16
Bufo terrestris: 12
Bufo cf. *Bufo terrestris:* 7
Bufo woodhousii fowleri: 1, 2, 4, 5-I, 5-II, 6, 8-II, 9, 10, 11, 17, 18-II, 18-III, 18-IV, 19
Bufo cf. *Bufo woodhousii:* 7
Bufo sp.: 3-I, 3-II, 3-III, 5-I, 5-II, 6, 7, 16
Hylidae
Hyla chrysoscelis or *versicolor:* 2, 12, 16
Hyla gratiosa: 5-I
Hyla sp.: 6, 17, 20
Pseudacris crucifer: 1, 2, 3-I, 7, 8-I, 8-II, 9, 11, 12, 16
Pseudacris triseriata: 1
Microhylidae
Gastrophryne carolinensis: 11, 12
Ranidae
Rana catesbeiana: 2, 4, 5-II, 9, 10, 11
Rana clamitans: 1, 2, 3-I, 4, 9, 10, 11
Rana (?) *clamitans:* 8-I
Rana palustris: 9
Rana (?) *palustris:* 8-I, 8-II, 10
Rana pipiens complex: 1, 2, 3-I, 3-II, 3-III, 4, 5-I, 5-II, 5-III, 6, 8-I, 8-II, 9, 10, 11, 13, 15, 16, 17, 18-II
Rana utricularia: 12
Rana sylvatica: 1, 2, 3-I, 3-II, 4, 8-I, 8-II, 9, 10, 15, 16, 19
Rana sp.: 5-III, 6, 7, 11, 16, 19
Kinosternidae
Kinosternon subrubrum: 12
Kinosternidae indet.: 11
Chelydridae
Chelydra serpentina: 10, 18-III, 18-IV
Emydidae
Chrysemys picta: 1, 5-III, 11, 12
(?) *Chrysemys picta:* 6, 18-I, 18-III
Clemmys insculpta: 7, 12, 14, 16
Clemmys muhlenbergii: 1
Emydoidea blandingii: 3-III, 14
Graptemys geographica: 5-I, 5-II, 9
Pseudemys concinna: 5-I, 5-II, 12
Pseudemys nelsoni: 12

Terrapene carolina carolina: 12
Terrapene carolina: 1, 4, 10, 11, 14, 17
(?) *Terrapene carolina:* 6
Trachemys scripta: 19, 20
Trachemys or *Pseudemys:* 4, 5-I, 11
Large emydidae indet.: 11
Emydidae indet.: 2

Testudinidae
Geochelone crassiscutata: 12
cf. *Geochelone* sp.: 14

Trionychidae
Apalone spinifera: 5-I
Apalone sp.: 5-I, 5-II, 13

Testudines indet.: 20

Phrynosomatidae
Sceloporus undulatus: 2, 3-I, 3-III, 6, 9, 10, 11, 12, 15
Sceloporus cf. *Sceloporus undulatus:* 1

Polychridae
Anolis carolinensis: 5-III, 11, 12

Scincidae
Eumeces fasciatus: 2, 7, 15, 19
Eumeces cf. *Eumeces fasciatus:* 1
Eumeces laticeps: 2, 3-I
Eumeces cf. *Eumeces laticeps:* 6
Eumeces (?) *laticeps:* 9
Eumeces sp.: 20
(?) *Eumeces* sp.: 4
Scincella lateralis: 10, 11

Colubridae
Carphophis amoenus: 1, 2, 3-I, 3-II, 3-III, 5-I, 5-II, 7, 8-I, 9, 10, 11, 12, 16, 17
Carphophis sp.: 20
Cemophora coccinea: 9, 10
Cemophora or *Lampropeltis:* 16
Coluber constrictor: 1, 2, 3-I, 3-III, 5-II, 15, 17
(?) *Coluber constrictor:* 12, 13
Coluber sp.: 21
Coluber or *Masticophis:* 4, 6, 8-I, 9, 10, 11, 14, 19
Diadophis punctatus: 1, 3-I, 3-II, 4, 5-II, 7, 8-I, 8-II, 9, 10, 11, 12, 16, 19
Elaphe guttata: 4, 6, 8-I, 9, 10, 11
Elaphe cf. *Elaphe guttata:* 5-I
Elaphe obsoleta: 4, 6, 8-I, 8-II, 9, 10, 11
Elaphe cf. *Elaphe obsoleta:* 18-II
Elaphe vulpina: 1, 3-III, 5-I, 5-II, 12, 13, 15
Elaphe sp.: 2, 3-I, 3-II, 3-III, 5-I, 5-II, 13, 20

(?) *Elaphe* sp.: 19
Heterodon platirhinos: 3-I, 3-II, 3-III, 9, 12, 13, 15, 16, 19
Heterodon cf. *Heterodon platirhinos:* 1, 5-I, 5-II, 5-III
(?) *Heterodon platirhinos:* 6
Lampropeltis calligaster rhombomaculata: 17
Lampropeltis calligaster: 4, 8-I, 9, 11, 12
Lampropeltis getula: 4, 5-I, 5-II, 5-III, 6, 7, 8-I, 8-II, 9, 10, 11, 12, 19
Lampropeltis triangulum triangulum: 5-I, 5-II, 5-III, 12
Lampropeltis triangulum: 1, 2, 3-I, 3-II, 3-III, 4, 6, 8-I, 9, 10, 11, 13, 15, 19
Lampropeltis sp.: 20
Nerodia (?) *erythrogaster:* 10
Nerodia sipedon: 1, 2, 3-I, 3-II, 4, 5-I, 5-II, 5-III, 6, 8-I, 8-II, 9, 10, 11, 12, 13, 15, 16, 18-II, 19
Nerodia sp.: 5-I, 5-II, 19
Opheodrys aestivus: 5-I, 5-II, 8-I, 9
Opheodrys (?) *aestivus:* 11
Opheodrys vernalis: 1, 4, 9, 10, 13
Opheodrys sp.: 7
Pituophis melanoleucus: 8-I, 8-II, 11
Pituophis sp.: 21
(?) *Regina septemvittata:* 8-I, 9, 10
Rhadinaea flavilata: 7
Storeria dekayi: 7, 8-I, 8-II, 9, 10, 11, 12
Storeria occipitomaculata: 8-I, 9, 10
Storeria (?) *occipitomaculata:* 4
Storeria sp.: 1, 2, 3-I, 3-II, 3-III, 5-I, 5-II, 7, 18-II
(?) *Storeria* sp.: 11
Thamnophis brachystoma: 3-I, 3-II, 3-III, 16
Thamnophis (?) *sauritus:* 8-I, 9, 11
Thamnophis sirtalis: 3-I, 3-II, 3-III, 4, 5-I, 5-II, 5-III, 6, 7, 8-I, 8-II, 9, 10, 11, 12, 16, 17, 19
Thamnophis sp.: 1, 2, 3-I, 3-II, 3-III, 5-I, 5-II, 5-III, 8-I, 8-II, 10, 13, 15, 18-II, 19, 20
Virginia striatula: 9, 10, 11
Virginia valeriae: 8-I, 8-II, 9, 10, 11, 16
(?) *Virginia valeriae:* 6
Virginia sp.: 7

Viperidae
Agkistrodon contortrix: 3-III, 4, 8-I, 8-II, 10, 11, 15, 16, 17
Agkistrodon piscivorous: 12
Agkistrodon sp.: 5-III
Crotalus horridus: 1, 2, 3-I, 3-III, 4, 5-I, 5-II, 5-III, 6, 8-I, 8-II, 9, 10, 12, 13, 15, 17, 19
Crotalus sp.: 21

Sistrurus catenatus: 15
Sistrurus or *Crotalus:* 16

Southeastern Region

The Southeastern region contains an abundance of cave, fissure fill, and spring run sites (most of them unstratified), as well as coastal marsh Pleistocene sites. Florida contains by far the largest number of Pleistocene herpetofaunas of any state in the United States or province of Canada.

Special Notes. A very large form of the eastern box turtle, *Terrapene carolina*, called *Terrapene carolina putnami* (Fig. 42A,E) occurs in the North American Pleistocene and has been widely discussed in Florida (e.g., Auffenberg, 1958; Milstead, 1967). Moreover, at least three modern subspecies of *Terrapene carolina* as well as *Terrapene carolina putnami* are represented individually or by various hybrid populations at different sites. The situation is so complicated that I have decided to list all of these forms as "*Terrapene carolina* ssp." and the remaining accounts where the large form occurs. This subspecies complex will be addressed in Chapter 8.

Many southeastern Rancholabrean sites have one or two extinct land tortoise species, a large ("giant") form, ***Geochelone crassiscutata*** (Fig. 43A), and a smaller form, ***Geochelone incisa*** (Fig. 43B). The giant form, especially, is of considerable interest, as it probably played the role of a large herbivore in the Pleistocene; like many of the large mammalian herbivores, it became extinct at the end of the epoch. These forms will also be adressed in Chapter 8.

The following states, mainly in the coastal plain of the Deep South, have yielded important Pleistocene herpetofaunas (Fig. 29):

South Carolina
Florida
Mississippi

Sites of Special Interest

1. Inglis IA Site, Florida

The Inglis IA Site of Citrus County in southwestern peninsular Florida is of considerable importance because it has yielded the largest Irvingtonian Land Mammal Age herpetological deposit in the Southeastern region. The fossils of this deposit accumulated in a sinkhole. Based on faunal evidence, the site is considered to be of the earliest Irvingtonian age, IrI (Meylan, 1982). It is possible that someday this site might be moved back into the Blancan Land Mammal Age (latest Pliocene). Nevertheless, this locality gives us a marvellous window on the herpetological life in Florida near the onset of Pleistocene times.

Inglis IA contains not only extralimital western (cf. *Gerrhonotus*, *Heterodon nasicus*) and northern (*Ambystoma* sp., *Eumeces* cf. *Eumeces fasciatus*,

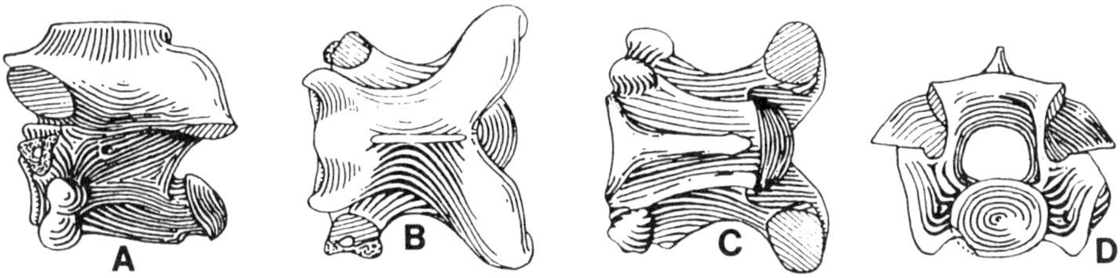

FIGURE 31. Holotype vertebra (University of Florida, Florida State Museum Number 6413) of *Diadophis elinorae* Auffenberg from the Pliocene of Haile, Alachua County, Florida. (A) Lateral, (B) dorsal, (C) ventral, (D) anterior [from Auffenberg, 1963b]. This species has also been reported from the Irvingtonian Inglis IA site, Citrus County, Florida, by Meylan (1982).

Opheodrys ventralis) forms, but extinct species of squamates. It is the type locality of *Eumeces carri* Meylan, 1982, an extinct lizard of the family Scincidae that differs from modern species of *Eumeces* in features of the teeth and dentary bone (Meylan, 1982, p. 18 and Fig. 1G–J).

It is also the type locality of *Regina intermedia* Meylan, 1982, an extinct crayfish snake of the family Colubridae, that differs from other species of the genus on the basis of features of the dentary bone (Meylan, 1982, pp. 40–41 and Fig. 10). *Diadophis elinorae* Auffenberg, 1963b (Fig. 31 in this book) was originally described from the Pliocene of the Haile locality in Alachua County, Florida. It differs from extant species of *Diadophis* in several vertebral features (Auffenberg, 1963b).

References. Meylan (1982, 1984).

Herpetofauna. *Ambystoma* sp., *Scaphiopus* sp., *Bufo* sp., *Hyla cinerea*, *Gastrophryne* sp., *Rana areolata*, *Rana catesbeiana*, *Chelydra* sp., *Terrapene* sp., *Trachemys scripta*, *Geochelone crassiscutata*, *Geochelone incisa*, *Gopherus* sp., *Apalone* sp., *Sceloporus undulatus*, *Eumeces carri* **(type locality)**, *Eumeces* cf. *Eumeces fasciatus*, cf. *Gerrhonotus*, *Ophisaurus ventralis*, cf. *Rhineura floridana*, *Cemophora coccinea*, *Coluber constrictor*, *Diadophis elinorae*, *Drymarchon corais*, *Elaphe guttata*, *Elaphe obsoleta*, *Farancia abacura*, *Heterodon nasicus*, *Heterodon platirhinos*, *Lampropeltis getula*, *Lampropeltis triangulum*, *Masticophis flagellum*, *Nerodia cyclopion*, *Nerodia erythrogaster*, *Opheodrys vernalis*, *Pituophis melanoleucus*, *Regina intermedia* **(type locality)**, *Rhadinaea* cf. *Rhadinaea flavilata*, *Stilosoma extenuatum*, *Tantilla* sp., *Thamnophis* cf. *Thamnophis sauritus*, *Virginia* sp., *Micrurus* cf. *Micrurus fulvius*, *Crotalus adamanteus*, *Sistrurus miliarius*, and *Alligator mississippiensis*.

2. Leisey Shell Pit Fauna, Florida

The Leisey Shell Pit fauna of Hillsborough County in southwestern peninsular Florida is the only other published early Irvingtonian Land Mammal Age herpetofauna from the Southeast region. The fossils at this site accumulated in a shell pit that contains marine, estuarine, freshwater, and terrestrial animals.

It has been suggested that the Leisey Shell Pit fauna represents Irvingtonian I, ca. 1.5 to 1.0 million ybp, based on the evolutionary stages of the mammals recovered from the site and thus represents the late part of the early Irvingtonian (Hulbert and Morgan, 1989). The known herpetofauna is mainly similar to one that could be found in the area today except for two extinct tortoises and the extralimital alligator snapper, *Macroclemys temminckii*. Sea turtle remains that were found at this site were not included in the herpetofaunal list that follows.

Reference. Hulbert and Morgan (1989)

Herpetofauna. *Siren* sp., *Bufo terrestris*, cf. *Rana*, *Kinosternon* sp., *Macroclemys temminckii*, *Terrapene carolina* spp., *Trachemys scripta*, *Geochelone crassiscutata*, *Geochelone incisa* complex, *Gopherus polyphemus*, *Apalone ferox*, cf. *Coluber*, *Elaphe* sp., *Farancia* sp., *Nerodia* sp., *Agkistrodon piscivorous*, *Crotalus adamanteus* and *Alligator mississippiensis*.

3. Reddick I Site, Florida

The Reddick I site in Marion County in north-central Florida is the richest, most diverse, and one of the most well-studied Pleistocene herpetofaunas in North America. The site is a complex cave and fissure fill complex that contains the so-called rodent beds that were derived mainly from the pellets cast by barn owls (*Tyto alba*). The age of the site is Rancholabrean and it may represent the Sangamonian age (Webb, 1974).

At least fifty herpetological taxa have been reported from the Reddick I site. Many reports have listed the herpetological remains of this site, but the standard reference is still Gut and Ray (1963). These authors provide a concise discussion of the geology of the site and divide it into localities A, B, C, and D. The herpetofauna listed below is from a composite of these localities. The Reddick herpetofauna is similar to one that may be found in the area today except for two extinct species of tortoises, an extinct species of eellike salamander (*Pseudobranchus robustus*, Fig. 33), an extinct subspecies of slider turtles (*Trachemys scripta petrolei*), and three extralimital northern species (*Bufo woodhousii fowleri*, *Eumeces* cf. *Eumeces fasciatus*, and *Carphophis amoenus*).

References. Gut and Ray (1963), Auffenberg (1955, 1956, 1958, 1963b), Tihen (1962), Weaver and Robertson (1967), Wilson (1975), Holman (1976).

Herpetofauna. *Pseudobranchus robustus*, *Siren lacertina*, *Ambystoma tigrinum*, *Ambystoma* sp., *Scaphiopus holbrookii*, *Bufo quercicus*, *Bufo terrestris*, *Bufo woodhousii fowleri*, *Hyla cinerea*, *Hyla* sp., *Pseudacris nigrita*, *Gastrophryne carolinensis*, *Rana areolata*, *Rana grylio*, *Rana utricularia*, *Pseudemys* or *Trachemys*, *Terrapene carolina* ssp., *Trachemys scripta petrolei*, *Geochelone crassiscutata*, *Geochelone incisa*, *Gopherus polyphemus*, *Apalone ferox*, *Anolis carolinensis*, *Cnemidophorus sexlineatus*, *Eumeces* cf. *Eumeces fasciatus*, *Eumeces* sp., *Ophisaurus ventralis*, *Rhineura floridana*, *Carphophis amoenus*, *Coluber constrictor*, *Diadophis punctatus*, *Drymarchon corais*, *Elaphe guttata*, *Elaphe obsoleta*, *Elaphe* sp., *Farancia* sp. *Heterodon platirhinos*, *Heterodon simus*, *Lampropeltis getula*, *Masticophis flagellum*, *Opheodrys aestivus*, *Pituophis melanoleucus*, *Rhadinaea flavilata*, *Storeria* cf. *Storeria dekayi*,

Tantilla relicta, *Thamnophis sirtalis*, *Micrurus fulvius*, *Crotalus adamanteus*, *Sistrurus miliarius*, and *Alligator mississippiensis*.

4. Williston IIIA Site, Florida

The Williston IIIA site of Levy County in north-central Florida consists of a single fossiliferous zone in one small sinkhole or "pipe" and probably represents much more of a "unit fauna" than many other major Florida herpetological sites. I excavated the entire site in the summer of 1956 and recently restudied some of the unidentified herpetological material, adding to the species list in Holman (1959a). The site is Rancholabrean and has been correlated with the Sangamonian interglacial age (Webb, 1974). Except for the presence of an extinct species of tortoise (*Geochelone incisa,* reported for the first time here), all the amphibians and reptiles of the deposit represent extant taxa that could be found living at or near the locality today.

References. Holman (1959a), Tihen (1962), Auffenberg (1963b).

Herpetofauna. *Ambystoma tigrinum*, *Ambystoma* sp., *Plethodon glutinosus* complex, *Scaphiopus holbrookii*, *Bufo terrestris*, *Hyla femoralis*, *Hyla* sp., *Gastrophryne carolinensis*, *Rana utricularia*, *Rana* sp., *Pseudemys floridana*, *Terrapene carolina* ssp., *Geochelone incisa*, *Gopherus polyphemus*, *Apalone ferox*, *Anolis carolinensis*, *Eumeces* cf. *Eumeces inexpectatus*, *Ophisaurus ventralis*, *Ophisaurus* sp., *Rhineura floridana*, *Coluber constrictor*, *Diadophis punctatus*, *Drymarchon corais*, *Elaphe* sp., *Heterodon platirhinos*, *Heterodon simus*, *Masticophis flagellum*, *Pituophis melanoleucus*, *Rhadinaea flavilata*, *Thamnophis* cf. *Thamnophis sirtalis*, *Thamnophis* sp., *Micrurus fulvius*, *Agkistrodon piscivorous*, *Crotalus adamanteus*, *Sistrurus miliarius*, and *Alligator mississippiensis*.

5. Arredondo Site, Florida

The Arredondo site, a sinkhole and fissure fill complex, is in Alacua County in north-central Florida. Several localities have been designated at Arredondo, but the main herpetological ones are termed Arredondo IA and IIA. A very large herpetofauna has arisen from collections at Arredondo over the years, and several authors have contributed to the knowledge of herpetological fossils from the site. The amphibian fauna is especially well-developed at Arredondo. The Arredondo complex is Rancholabrean and is considered to represent a time period ranging from Sangamonian to Wisconsinan (Webb, 1974; Lundelius et al., 1983).

Many of the amphibians and reptiles represent extant taxa that occur in the area today; but an extinct species of eellike salamander, *Pseudobranchus* cf. *Pseudobranchus robustus* (Fig. 33), an extinct subspecies of slider turtle, *Pseudemys scripta petrolei*, and an extinct species of small tortoise, *Geochelone incisa*, occur there. Arredondo is the type locality of *Hyla baderi* Lynch, 1965, an extinct species of treefrog that differs from other species of the genus on the basis of ilial characters (Fig. 32). Extralimital northern taxa, a mudpuppy,

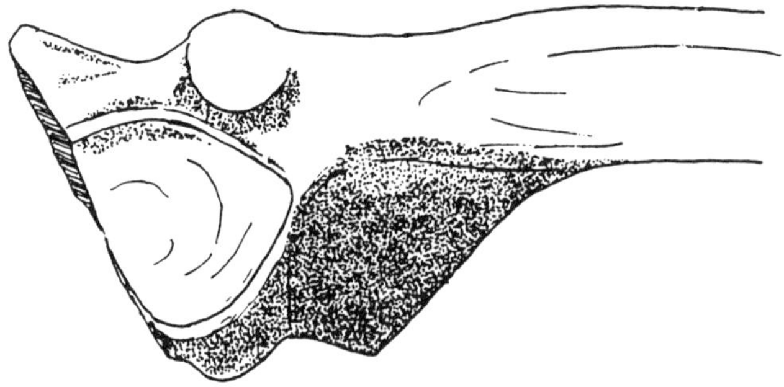

FIGURE 32. Holotype right ilium (University of Florida, Florida State Museum Number 9113) of *Hyla baderi* Lynch from the Rancholabrean Arredondo locality of Alachua County, Florida. [From Lynch, 1965]

Necturus, and a small snake, *Carphophis amoenus*, also occur at the Arredondo site.

References. Auffenberg (1956, 1958, 1963a, 1963b), Holman (1962a), Tihen (1962), Lynch (1965), Weaver and Robertson (1967).

Herpetofauna. *Pseudobranchus* cf. *Pseudobranchus robustus*, *Siren lacertina*, *Notophthalmus viridescens*, *Necturus* sp., *Ambystoma tigrinum*, *Scaphiopus holbrookii*, *Bufo terrestris*, *Acris gryllus*, *Hyla baderi* **(type locality)**, *Hyla cinerea*, *Hyla gratiosa*, *Pseudacris ornata*, *Gastrophryne carolinensis*, *Rana catesbeiana*, *Rana utricularia*, *Rana* sp., *Terrapene carolina* ssp., *Trachemys scripta petrolei*, *Geochelone incisa*, *Anolis carolinensis*, *Carphophis amoenus*, *Cemophora coccinea*, *Coluber constrictor*, *Drymarchon corais*, *Elaphe guttata*, *Elaphe obsoleta*, *Heterodon platirhinos*, *Lampropeltis getula*, *Masticophis flagellum*, *Nerodia fasciata*, *Opheodrys aestivus*, *Stilosoma extenuatum*, *Storeria dekayi*, *Tantilla relicta*, *Thamnophis sirtalis*, *Micrurus fulvius*, and *Crotalus adamanteus*.

6. Haile (Rancholabrean) Site, Florida

Herpetological sites of Rancholabrean age occur near the village of Haile in Alachua County in north-central Florida, and a very substantial list of Rancholabrean amphibian and reptile species at Haile has built up over the years. The fossils occur in a complex of sinkholes and fissures, and at least twelve localities in the Rancholabrean portion of the site have been named. The Haile Rancholabrean complex is thought to represent a time period ranging from Sangamonian into Wisconsinan times (Webb, 1974; Lundelius et al., 1983).

The herpetofauna mainly represents species that are extant and occur in the area today. Two extinct species of tortoise and an extinct species of eellike salamander are exceptions. Haile is the type locality of *Pseudobranchus robustus* Goin and Auffenberg, 1955, which was described on the basis of a middle thoracic vertebra (Fig. 33). The validity of this extinct species has been questioned and the subject will be further addressed in Chapter 8. Extralimital northern species *Bufo woodhousii fowleri* and *Carphophis amoenus* also occur at Haile.

One of the significant features of the site is a succession of changes of subspecies and subspecies intergrades of *Terrapene carolina* from the lower to the upper strata of the site.

References. Auffenberg (1955, 1956, 1963a, 1963b, 1967), Goin and Auffenberg (1955), Tihen (1952, 1962), Lynch (1964).

Herpetofauna. *Pseudobranchus robustus* **(type locality)**, *Bufo terrestris*, *Bufo woodhousii fowleri*, *Bufo* sp., *Hyla femoralis*, *Hyla gratiosa*, *Hyla squirella*, *Rana grylio*, *Terrapene carolina* ssp., *Geochelone crassiscutata*, *Geochelone*

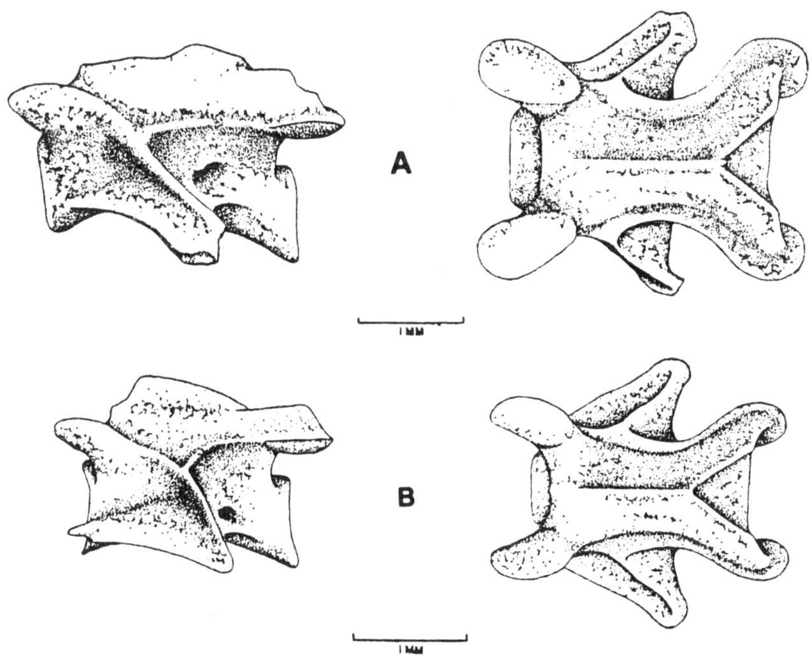

FIGURE 33. (A) holotype thoracic vertebra (Museum of Comparative Zoology, Harvard, Number 2279) of *Pseudobranchus robustus* Goin and Auffenberg from the Rancholabrean of Haile, Alachua County, Florida. (B) Thoracic vertebra from a modern skeleton of *Pseudobranchus striatus*. [From Goin and Auffenberg, 1955]

incisa, *Anolis carolinensis*, *Ophisaurus compressus*, *Ophisaurus ventralis*, *Carphophis amoenus*, *Coluber constrictor*, *Drymarchon corais*, *Elaphe guttata*, *Elaphe obsoleta*, *Heterodon platirhinos*, *Heterodon simus*, *Lampropeltis getula*, *Masticophis flagellum*, *Nerodia cyclopion*, *Nerodia* cf. *Nerodia erythrogaster*, *Nerodia fasciata*, *Opheodrys aestivus*, *Thamnophis sirtalis*, *Micrurus fulvius*, *Agkistrodon piscivorous*, *Crotalus adamanteus*, and *Sistrurus miliarius*.

7. Edisto Beach Site, South Carolina

The Edisto Beach site of Colleton County in extreme southern coastal South Carolina is thus far the only significant published herpetological site in the Carolinas. The fauna accumulated as an ocean beach deposit. The site is considered to represent a very late Wisconsinan age of from about 12,000 to about 10,000 ybp (Lundelius et al., 1983). Other than an extinct tortoise and an extinct subspecies of slider turtle, *Trachemys scripta* cf. *petrolei*, the fauna represents taxa that could be expected to occur in the area today. A record of cf. *Gavialosuchus* from this site, reported by Roth and Laerm (1980), has been omitted from the list below (see discussion in Holman, 1991b).

References. Dobie and Jackson (1979), Roth and Laerm (1980), Holman (1991b).

Herpetofauna. *Kinosternon* sp., *Chelydra serpentina*, *Malaclemys terrapin*, *Pseudemys concinna* or *floridana*, *Pseudemys floridana, Pseudemys nelsoni*, *Terrapene carolina* ssp., *Trachemys scripta* cf. *petrolei*, *Geochelone* sp., *Gopherus* sp., *Apalone* sp., cf. *Coluber* or *Masticophis*, cf. *Elaphe* sp., and *Alligator mississippiensis*.

8. Devil's Den Chamber 3, Florida

The Devil's Den Chamber 3 herpetofauna in Levy County in north-central Florida is of importance because is represents a sinkhole trap or cenote where vertebrates accumulated in the very late Wisconsinan. It has been suggested that the accumulation occurred about 10,000 ybp, but it is possible that some of the vertebrates in the site represent a later time. The fauna contains two species of extinct tortoises, one of which could be a dwarf form of the giant tortoise *Geochelone crassiscutata*, and one intrusive northern species *Bufo woodhousii fowleri*. The rest of this rather large herpetofauna consists of extant taxa that might be expected to be found living near the area today.

Reference. Holman (1978).

Herpetofauna. *Ambystoma tigrinum*, *Scaphiopus holbrookii*, *Bufo terrestris*, *Bufo woodhousii fowleri*, *Rana catesbeiana*, *Rana utricularia*, *Kinosternon subrubrum*, *Chelydra serpentina*, *Deirochelys reticularia*, *Terrapene carolina* ssp., *Trachemys scripta*, *Geochelone crassiscutata*, *Geochelone incisa*, *Gopherus polyphemus*, *Anolis carolinensis*, *Coluber* or *Masticophis*, *Drymarchon corais*, *Elaphe guttata*, *Elaphe obsoleta*, *Farancia* sp., *Heterodon platirhinos*, *Heterodon simus*, *Lampropeltis getula*, *Lampropeltis triangulum*, *Nerodia erythrogaster* or *fasciata*, *Pituophis melanoleucus*, *Thamnophis sirtalis*, *Micrurus fulvius*, *Agkistrodon piscivorous*, *Crotalus admamanteus*, and *Alligator mississippiensis*.

9. Sabertooth Cave, Florida

The Sabertooth Cave site lies near the town of Lecanto in Citrus County in the northwestern part of peninsular Florida. The herpetofauna that is reported upon here is of importance because it was collected in one afternoon from a narrow stratigraphic zone, and thus is thought to represent a rather narrow window of time. All of the fossils of the site are the result of a natural cave accumulation. The age of the deposit is attributed to the Wisconsinan based on the faunal assemblage (Webb, 1974). I believe that the site represents a time very late in the Wisconsinan. All the herpetological species represent extant taxa and all of them could be found living in Citrus County, Florida, today. Since my original report on this collection (Holman, 1958), I have reexamined some of the material and have made a few taxonomic changes.

Reference. Holman (1958).

Herpetofauna. *Plethodon glutinosus* complex, *Scaphiopus holbrookii*, *Gastrophryne carolinensis*, *Rana utricularia*, *Pseudemys* or *Trachemys*, *Terrapene carolina* ssp., *Gopherus polyphemus*, *Anolis carolinensis*, *Eumeces* sp., *Ophisaurus ventralis*, *Rhineura floridana*, *Carphophis amoenus*, *Coluber constrictor*, *Elaphe* sp., *Heterodon platirhinos*, *Masticophis flagellum*, *Nerodia fasciata*, *Rhadinaea flavilata*, *Micrurus fulvius*, *Agkistrodon piscivorous*, and *Crotalus adamanteus*.

10. Ichetucknee River Site, Florida

The Ichetucknee River site in Columbia County in the far northern central part of the Florida peninsula is an important fluvial spring run site that has yielded a large herpetofauna. This site probably represents the latest Wisconsinan, but there is a chance that some younger vertebrate material is mixed in. Nevertheless, two extinct tortoise species and an extinct slider turtle subspecies, *Trachemys scripta petrolei*, come from the site and one northern invader, *Bufo woodhousii fowleri*, is known.

References. Auffenberg (1956, 1958, 1963a, 1963b), Tihen (1962), Weaver and Robertson (1967)

Herpetofauna. *Bufo terrestris*, *Bufo woodhousii fowleri*, *Terrapene carolina* ssp., *Trachemys scripta petrolei*, *Geochelone crassiscutata*, *Geochelone incisa*, *Anolis carolinensis*, *Coluber constrictor*, *Drymarchon corais*, *Elaphe guttata*, *Elaphe obsoleta*, *Farancia* sp., *Heterodon platirhinos*, *Lampropeltis getula*, *Nerodia cyclopion*, *Nerodia fasciata*, *Nerodia taxispilota*, *Pituophis melanoleucus*, *Regina alleni*, *Thamnophis sirtalis*, and *Crotalus adamanteus*.

11. Vero Beach Strata 2 and 3, Florida

The Vero Beach site in Indian River County in southeastern coastal Florida is the most important coastal marsh Pleistocene herpetofauna in the Southeastern region. The age of Strata 2 and 3 are believed to be late Wisconsinan based on the faunal assemblage (Webb, 1974). This site, which has been known for a long time, was most thoroughly and systematically excavated by Weigel (1962). The site contains an extinct tortoise as well as the extralimital northern forms

Ambystoma sp. and *Sternotherus minor*, but the majority of the herpetofauna could be found in the area today. Marine turtles reported from the site by Weigel (1962) are not included here. I have made a few taxonomic changes to some of the original papers.

References. Brattstrom (1953c, 1954b), Auffenberg (1956, 1958), Tihen (1962), Weigel (1962), Jackson (1978).

Herpetofauna. *Siren lacertina*, *Amphiuma means*, *Ambystoma* sp., *Scaphiopus holbrookii*, *Bufo terrestris*, *Bufo* sp., *Rana* cf. *Rana catesbeiana*, *Rana utricularia*, *Kinosternon baurii*, *Kinosternon subrubrum*, *Sternotherus minor*, *Sternotherus odoratus*, *Chelydra serpentina*, *Deirochelys reticularia*, *Pseudemys* cf. *Pseudemys floridana*, *Terrapene carolina* spp., *Trachemys scripta*, *Geochelone crassiscutata*, *Gopherus polyphemus*, *Anolis carolinensis*, *Sceloporus undulatus*, *Eumeces* sp., *Ophisaurus compressus*, *Coluber constrictor*, *Coluber* or *Masticophis*, *Drymarchon corais*, *Farancia abacura*, *Farancia* sp., *Lampropeltis getula*, *Nerodia fasciata*, *Nerodia taxispilota*, *Nerodia* sp., *Pituophis melanoleucus*, *Thamnophis* sp., *Micrurus fulvius*, *Crotalus adamanteus*, and *Alligator mississippiensis*.

12. Little Salt Spring Site, Florida

The Little Salt Spring site from Sarasota County in coastal southwestern peninsular Florida is of exceptional interest because a giant tortoise found there, *Geochelone crassiscutata*, was either killed by or manipulated over a fire by a wooden human spear. The vertebrate fauna of the Little Salt Spring site accumulated on a ledge in a large sinkhole pond when the water table in southwestern Florida was markedly lower than it is today. A radiocarbon date of 12,030 $\pm$ 200 ybp was obtained on the wooden spear found between the plastron and carapace of the extinct giant tortoise.

References. Clausen et al. (1979), Holman and Clausen (1984).

Herpetofauna. *Pseudemys floridana*, *Pseudemys nelsoni*, *Terrapene carolina* ssp., *Geochelone crassiscutata*, *Gopherus polyphemus*, and *Crotalus adamanteus*.

Other Important Sites

13. Haile XVIA Site, Alachua County, Florida

The fossil vertebrates of this site accumulated in one of the Haile, Florida, limestone solution features. This is an early Irvingtonian site (IrI) and the herpetofauna is said to be very similar to the earlier Irvingtonian site at Inglis IA, Florida, (site 1 this section). The fauna has not been published in its entirity.

Reference. Meylan (1984).

14. Coleman IIA, Sumpter County, Florida

The herpetofauna at this site accumulated among an extensive sinkhole assemblage of vertebrates. This is also an Irvingtonian site whose herpetofauna is said to be quite similar to the earlier Irvingtonian site at Inglis IA, Florida (site I this section) and also has not been published in its entirity. The Coleman IIA site is noteworthy as the type locality for the most recently described extinct North American small Pleistocene tortoise, *Geochelone mlynarskii* Auffenberg, 1988. This species is most closely related to the small Rancholabrean species *Geochelone incisa*.

References. Meylan (1984), Auffenberg (1988).

15. Payne's Prairie Site, Alachua County, Florida

This site was recovered from spoil along the excavation of a canal and was primarily deposited in a very large "dish"-type sinkhole. It is thought to be of Sangamonian age on the basis of mammalian evidence (Webb, 1974).

Reference. Auffenberg (1963b).

16. Orange Lake Site, Marion County, Florida

The Orange Lake site was also deposited in a very large "dish"-type sinkhole and is also thought to be Sangamonian in age based on its fauna and stratigraphic position. I have somewhat modified the taxonomy of my original paper on the site.

Reference. Holman (1959b).

17. Reddick II Site, Marion County, Florida

This site is distinct from the localities attributed to the Reddick I site (site 3 this section), but it also developed as a limestone solution feature complex. It is considered to represent the Sangamonian age (Webb, 1974).

References. Auffenberg (1958, 1963a), Jackson (1978).

18. Rock Springs Site, Orange County, Florida

This site is the result of spring run erosion of Pleistocene sediments. It is considered to represent the Sangamonian age (Webb, 1974).

References. Auffenberg (1956, 1963a, 1963b).

19. Bradenton Site, Manatee County, Florida

This site is a coastal marsh locality that is considered to represent the Sangamonian age based on the vertebrate fauna (Webb, 1974).

References. Auffenberg (1956, 1958, 1963b), Tihen (1962), Jackson (1978).

20. Haile XIVA Site, Alachua County, Florida

This site is distinct from the other complex of Haile, Florida, localities (site 6 this section). This site occurs in a sinkhole fissure fill. It probably represents a transitional stage between the late Sangamonian and early Wisconsinan.

Reference. Martin, R. (1974).

21. Santa Fe River Site I, Columbia–Gilchrist County Line, Florida

This site results from Pleistocene sediments eroded by the Santa Fe River. It is considered to be a heterochronic Blancan–Rancholabrean assemblage. The turtle remains listed in the taxonomic summary here are reported to have had the "coloration" of Rancholabrean fossils (Jackson, 1975).

Reference. Jackson (1975).

22. Kanapaha I Site, Alachua County, Florida

This site consists of a complex of solution channels and fissure fills. It is believed to be a Rancholabrean complex.

References. Tihen (1962), Auffenberg (1963a, 1963b).

23. Hornsby Springs Site, Alachua County, Florida

The fauna of this site accumulated in a sinkhole. A radiocarbon date of 9,800 $\pm$ 270 ybp was derived from shell marl, indicateing a very late Wisconsinan or very early Holocene age (Lundelius et al., 1983).

References. Tihen (1962), Auffenberg (1963b).

24. Kendrick IA Site, Marion County, Florida

This site is another sinkhole and fissure fill deposit that is believed to represent the Wisconsinan based on the mammalian assemblage (Webb, 1974).

References. Tihen (1962), Auffenberg (1963a), Jackson (1978).

25. Mefford Cave I and II, Marion County, Florida

The fauna in this site is the result of a natural cave accumulation. It is believed to represent the late Wisconsinan on the basis of the faunal assemblage (Lundelius et al., 1983).

References. Auffenberg (1958, 1963a, 1963b).

26. Eichelberger Cave A and B, Marion County, Florida

This site represents a natural cave accumulation and is believed to represent the "late Wisconsinan ?" based on the faunal assemblage (Lundelius et al., 1983).

References. Auffenberg (1958, 1963a, 1963b).

27. Wakulla Springs Site, Wakulla County, Florida

This spring run site is believed to represent the late part of the Rancholabrean Land Mammal Age (Lundelius et al., 1983).

Reference. Brattstrom (1953c).

28. Melbourne Site, Brevard County, Florida

This site is the result of a natural accumulation of bones in a coastal marsh. It is believed to represent the Wisconsinan based on the faunal assemblage (Webb, 1974).

References. Gilmore (1938), Brattstrom (1953c, 1954b), Auffenberg (1958, 1963a).

29. Seminole Field, Pinellas County, Florida

This fauna accumulated as a natural accumulation in a coastal marsh setting. It has believed to represent the Wisconsinan (Webb, 1974).

References. Gilmore (1938), Brattstrom (1953c, 1954b), Auffenberg (1958), Jackson (1978).

30. Winter Beach Site, St Lucie County, Florida

The Winter Beach site represents another natural accumulation of bones in a coastal marsh. It is believed to represent the Wisconsinan on the basis of its faunal assemblage.

References. Tihen (1962), Auffenberg (1956, 1958, 1963b).

31. Blackbelt Site Complex, Northeastern Mississippi

This is complex of localities with individually poor stratigraphic control. They represent the Rancholabrean Land Mammal Age and are thought to represent both the Sangamonian and Wisconsinan ages. The important extralimital forms *Emydoidea blandingii* (northern) and *Geochelone crassiscutata* (southern) will be discussed later in this book.

References. Kaye (1974), Jackson and Kaye (1974a, 1974b, 1975), Daly (1992).

32. Ardis Local Fauna, Dorchester County, South Carolina

The depositional environment represents a fluvial sandfill in a limestone solution feature. The site represents the late Rancholabrean based on radiocarbon dates. This site has produced the first authentic record of *Clemmys guttata* from the Pleistocene.

Reference. Bentley and Knight (1994).

Summary of Herpetological Taxa*

Sirenidae
Pseudobranchus robustus: 3, 6
Pseudobranchus cf. *Pseudobranchus robustus:* 5
Siren lacertina: 3, 5, 11, 30
Siren sp.: 2
Salamandridae
Notophthalmus viridescens: 5
Notophthalmus sp.: 13
Proteidae
Necturus sp.: 5
Amphiumidae
Amphiuma means: 11, 27, 29

*Numerals in summary correspond to sites described in this section.

Ambystomatidae
Ambystoma tigrinum: 3, 4, 5, 8, 16
Ambystoma sp.: 1, 3, 4, 11, 16
Plethodontidae
Plethodon glutinosus complex: 4, 9, 16
Pelobatidae
Scaphiopus holbrookii: 3, 4, 5, 8, 9, 11, 16
Scaphiopus sp.: 1
Bufonidae
Bufo quercicus: 3
Bufo terrestris: 2, 3, 4, 5, 6, 8, 10, 11, 16, 19, 22, 23, 30
Bufo woodhousii fowleri: 3, 6, 8, 10, 19, 22, 24
Bufo sp.: 1, 6, 11, 20, 30
Hylidae
Acris gryllus: 5
Hyla baderi: 5
Hyla cinerea: 1, 3, 5
Hyla femoralis: 4, 6
Hyla gratiosa: 5, 6
Hyla squirella: 6
Hyla sp.: 3, 4
Pseudacris nigrita: 3
Pseudacris ornata: 5
Microhylidae
Gastrophryne carolinensis: 3, 4, 5, 9, 16
Gastrophryne sp.: 1
Ranidae
Rana areolata: 1, 3
Rana catesbeiana: 1, 5, 8
Rana cf. *Rana catesbeiana:* 11
Rana grylio: 3, 6
Rana utricularia: 3, 4, 5, 8, 9, 11
Rana sp.: 4, 5, 16, 20, 30
cf. *Rana:* 2
Kinosternidae
Kinosternon baurii: 11
Kinosternon subrubrum: 8, 11
Kinosternon sp.: 2, 7, 13, 30
Sternotherus minor: 11
Sternotherus odoratus: 11
Chelydridae
Chelydra serpentina: 7, 8, 11, 31
Chelydra sp.: 1
Macroclemys temminckii: 2, 31

Emydidae
  Chrysemys picta: 31
  Clemmys guttata: 32
  Deirochelys reticularia: 8, 11, 17, 19, 24, 29
  Deirochelys sp.: 13
  Emydoidea blandingii: 31
  Graptemys cf. *Graptemys barbouri:* 21
  Malaclemys terrapin: 7
  Pseudemys concinna: 21
  Pseudemys concinna or *floridana:* 7
  Pseudemys floridana: 4, 7, 12
  Pseudemys cf. *Pseudemys floridana:* 11
  Pseudemys nelsoni: 7, 12, 21, 31
  Pseudemys or *Trachemys:* 3, 9
  Terrapene carolina ssp.: 2, 3, 4, 5, 6, 7, 8, 9, 10, 11, 12, 16, 17, 19, 20, 24, 25, 26, 28, 29, 30, 31
  Terrapene sp.: 1
  Trachemys scripta petrolei: 3, 5, 10
  Trachemys scripta cf. *petrolei:* 7
  Trachemys scripta: 1, 2, 8, 11, 21, 31
Testudinidae
  Geochelone crassiscutata: 1, 2, 3, 6, 8, 10, 11, 12, 17, 18, 20, 22, 25, 26, 28, 31
  Geochelone incisa: 1, 3, 4, 5, 6, 8, 10, 16, 20
  Geochelone incisa complex: 2
  Geochelone mlynarskii: 14
  Geochelone sp.: 7
  Gopherus polyphemus: 2, 3, 4, 8, 9, 11, 12, 20
  Gopherus sp.: 1, 7
Trionychidae
  Apalone ferox: 2, 3, 4
  Apalone sp.: 1, 7
Polychridae
  Anolis carolinensis: 3, 4, 5, 6, 8, 9, 10, 11, 14, 16, 18, 19, 30
Phrynosomatidae
  Sceloporus undulatus: 1, 11
Teiidae
  Cnemidophorus sexlineatus: 3
  Cnemidophorus cf. *Cnemidophorus sexlineatus:* 20
  Cnemidophorus sp.: 13
Scincidae
  Eumeces carri: 1
  Eumeces cf. *Eumeces fasciatus:* 1, 3
  Eumeces inexpectatus: 14

Eumeces cf. *Eumeces inexpectatus:* 4
Eumeces sp.: 3, 9, 11
Anguidae
cf. *Gerrhonotus:* 1
Ophisaurus compressus: 6, 11
Ophisaurus ventralis: 1, 3, 4, 6, 9, 16, 30
Ophisaurus sp.: 4
Rhinuridae
Rhineura floridana: 3, 4, 9, 20
cf. *Rhineura floridana:* 1
Colubridae
Carphophis amoenus: 3, 5, 6, 9, 14, 25
Cemophora coccinea: 1, 5
Coluber constrictor: 1, 3, 4, 5, 6, 9, 10, 11, 15, 20, 26, 28, 29, 30
cf. *Coluber:* 2
Coluber or *Masticophis:* 8, 11
cf. *Coluber* or *Masticophis:* 7
Diadophis elinorae: 1
Diadophis punctatus: 3, 4, 25
Drymarchon corais: 1, 3, 4, 5, 6, 8, 10, 11, 15, 16, 18, 22, 28, 29
Elaphe guttata: 1, 3, 5, 6, 8, 10, 15, 25, 26
Elaphe cf. *Elaphe guttata:* 20
Elaphe obsoleta: 1, 3, 5, 6, 8, 10, 22, 29
Elaphe sp.: 2, 3, 4, 9, 16
cf. *Elaphe* sp.: 7
Farancia abacura: 1, 11, 29
Farancia sp.: 2, 3, 8, 10, 11, 16, 23, 30
Heterodon nasicus: 1
Heterodon platirhinos: 1, 3, 4, 5, 6, 8, 9, 10, 20, 25, 26
Heterodon simus: 3, 4, 6, 8, 14, 25
Lampropeltis getula: 1, 3, 5, 6, 8, 10, 11, 16, 20, 28, 29
Lampropeltis triangulum: 1, 8, 29
Lampropeltis cf. *Lampropeltis triangulum:* 20
Masticophis flagellum: 1, 3, 4, 5, 6, 9, 20, 26
Nerodia cyclopion: 1, 6, 10, 15, 19
Nerodia erythrogaster: 1
Nerodia cf. *Nerodia erythrogaster:* 6
Nerodia erythrogaster or *fasciata:* 8
Nerodia fasciata: 5, 6, 9, 10, 11, 19, 23, 30
Nerodia taxispilota: 10, 11, 23
Nerodia sp.: 2, 11, 27, 29
Opheodrys aestivus: 3, 5, 6, 22
Opheodrys vernalis: 1
Pituophis melanoleucus: 1, 3, 4, 8, 10, 11, 22, 26, 29
Regina alleni: 10, 30

Regina intermedia: 1
Rhadinaea flavilata: 3, 4, 9
Rhadinaea cf. *Rhadinaea flavilata:* 1
Stilosoma extenuatum: 1, 5
Stilosoma cf. *Stilosoma extenuatum:* 20
Storeria dekayi: 5
Storeria cf. *Storeria dekayi:* 3
Tantilla relicta: 3, 5
Tantilla sp.: 1
Thamnophis cf. *Thamnophis sauritus:* 1
Thamnophis sirtalis: 3, 5, 6, 8, 10, 15, 19, 23, 25, 30
Thamnophis cf. *Thamnophis sirtalis:* 4
Thamnophis sp.: 4, 11, 27, 29
Virginia sp.: 1

Elapidae
Micrurus fulvius: 3, 4, 5, 6, 8, 9, 11, 16, 20
Micrurus cf. *Micrurus fulvius:* 1

Viperidae
Agkistrodon piscivorous: 2, 4, 6, 8, 9, 28, 29
Crotalus adamanteus: 1, 2, 3, 4, 5, 6, 8, 9, 10, 11, 12, 16, 18, 23, 25, 26, 28, 29
Sistrurus miliarius: 1, 3, 4, 6, 20

Alligatoridae
Alligator mississippiensis: 1, 2, 3, 4, 7, 8, 11, 16

Ozark Region

The Ozark region is replete with caves and other solution features, yet it has yielded a smaller number of Pleistocene amphibian and reptile records than the solution features of the Appalachian and Southeastern regions. This is because of the restricted geographic area in which the Ozark uplift occurred. Only two states are included in this section (Fig. 29):

Missouri
Arkansas

Sites of Special Interest

1. Conard Fissure, Arkansas

The Conard Fissure in Newton County, Arkansas, is a classic site that was first documented by Brown (1908). The vertebrate fauna comes from a natural cave fill and was derived from various sources. The site is now considered to represent the Irvingtonian Land Mammal Age (IrII, Reppening, 1987) and thus is the only Irvingtonian site in the entire Ozark region. All the amphibians and reptiles from the site represent living taxa, and all occur in or near the area of the site today.

References. Brown (1908), Brattstrom (1954b), Dowling (1958), Tihen (1962).

Herpetofauna. *Ambystoma* sp., *Bufo woodhousii fowleri*, *Carphophis amoenus*, *Coluber constrictor*, *Elaphe guttata*, *Elaphe vulpina*, *Elaphe* sp., *Heterodon* sp., *Lampropeltis* (?) *calligaster*, *Lampropeltis getula*, *Lampropeltis triangulum*, *Masticophis flagellum*, *Pituophis melanoleucus*, *Thamnophis sirtalis*, and *Crotalus horridus*.

2. Boney Spring, Missouri

The Boney Spring late Wisconsinan fauna in Benton County in the western Ozark highlands of Missouri contains a moderately large herpetofauna that is probably older than those of the remaining Ozark sites. Evidence indicates that the accumulation occurred around the outlet of an artesian spring. Radiocarbon dates on organic matter indicate that the accumulation occurred between 16,200 and 13,600 ybp. All the herpetological remains represent living taxa and all could be found living in the area today.

Reference. Saunders (1977).

Herpetofauna. *Ambystoma opacum*, *Bufo* sp., *Rana catesbeiana*, *Rana* sp., *Eumeces* cf. *Eumeces fasciatus*, *Carphophis amoenus*, *Diadophis punctatus*, *Lampropeltis triangulum*, *Storeria* sp., *Thamnophis proximus* or *sirtalis*, and *Thamnophis* sp.

3. Brynjulfson Cave I, Missouri

This very late Wisconsinan or earliest Holocene site occurs in Boone County, Missouri, near Columbia. The vertebrate fauna was derived from a natural cave accumulation. Bone from the cave yielded a radiocarbon date of 9,440 ± 760 ybp. All the herpetological taxa are extant, but *Emydoidea blandingii* occurs south of its presumed range in Missouri today.

Reference. Parmalee and Oesch (1972).

Herpetofauna. *Bufo* sp., *Rana catesbeiana*, *Rana* sp., *Chelydra serpentina*, *Emydoidea blandingii*, *Terrapene* cf. *Terrapene ornata*, *Terrapene* sp., Lacertilia sp., Colubridae sp., Viperidae sp., and Serpentes sp.

4. Crankshaft Cave, Missouri

The Crankshaft Cave occurs in Jefferson County in southeastern Missouri. It has yielded a large herpetofauna, as Ozark Pleistocene herpetofaunas go, but there is a possibility that some of these remains may be early Holocene. The bones represent a natural cave accumulation. In general, based on the extinct mammals present, one may assume a very late Wisconsinan age for most of the vertebrate remains. None of the herpetological remains represent extinct taxa, and all of them could be found living in or near the area today.

Reference. Parmalee, Oesch, and Guilday (1969).

Herpetofauna. *Ambystoma maculatum*, *Ambystoma tigrinum*, *Eurycea* sp.,

Bufo sp., *Rana* sp., *Terrapene* cf. *Terrapene ornata*, *Sceloporus undulatus*, *Ophisaurus ventralis*, *Coluber constrictor* or *Masticophis flagellum*, *Elaphe vulpina*, *Elaphe* sp., *Heterodon platirhinos*, *Lampropeltis getula*, *Lampropeltis triangulum*, *Lampropeltis* sp., *Pituophis melanoleucus*, *Thamnophis proximus*, *Thamnophis sirtalis*, and *Crotalus horridus*.

5. Peccary Cave, Arkansas

Peccary Cave in Newton County, northwestern Arkansas, is important in having a relatively large number of amphibian fossils. The herpetofauna of this site is a part of a natural cave accumulation. The site is believed to be late Wisconsinan on the basis of peccary and musk ox remains (Davis, 1973). All the herpetological taxa identified are extant and all of them may be found living in or near the area today.

Reference. Davis (1973).

Herpetofauna. *Necturus* cf. *Necturus maculosus*, *Ambystoma maculatum*, *Ambystoma* cf. *Ambystoma tigrinum*, *Ambystoma* sp., *Eurycea* sp., *Plethodon* cf. *Plethodon dorsalis*, *Plethodon* cf. *Plethodon glutinosus* complex, *Scaphiopus* cf. *Scaphiopus holbrookii*, *Bufo americanus americanus*, *Rana catesbeiana*, *Rana* cf. *Rana palustris*, *Rana pipiens* complex, *Rana* sp., *Crotaphytus* sp., *Sceloporus* sp., *Cnemidophorus* cf. *Cnemidophorus sexlineatus*, *Eumeces* sp., and *Ophisaurus attenuatus*.

Other Important Sites

6. Jones Spring Fauna, Hickory County, Missouri

This fauna occurred in a peat deposit. Radiocarbon dates on peat indicated an age of greater than 40,000 ybp and the site is thought to represent the middle Wisconsinan (Moodie and Van Devender, 1977). The isolated occurrence of a large box turtle (*Terrapene carolina* ssp.) at Jones Springs is noteworthy.

Reference. Moodie and Van Devender (1977).

7. Crevice Cave, Perry County, Missouri

The fauna is the result of a natural accumulation in a cave and predators may have dragged some of the animals in. The site is believed to represent the "late Pleistocene" based on the fauna, which included jaguars, sabertooths, mastodonts, mammoths, horses, peccaries and bison. Of significance is the reported occurrence of "cf. *Geochelone*," an extinct tortoise.

Reference. Hawksley (1986).

8. Tom Moore Cave, Perry County, Missouri

This is another natural cave assemblage that is believed to represent the "late Pleistocene" on the basis of its faunal assemblage. This is the second Ozark region locality that has yielded a fossil interpreted as "cf. *Geochelone,*" an extinct tortoise.

Reference. Hawksley (1986).

9. Herculaneum Crevice, Jefferson County, Missouri

This fauna is a natural accumulation in a crevice. It is believed to represent the "late Pleistocene" based on the fauna. Scutes from the crevice have been interpreted as being from an alligator (*Alligator mississippiensis*). This occurrence needs confirmation.

Reference. Mehl (1962).

10. Enon Sink Local Fauna, Moniteau County, Missouri

This site represents a natural sink filling. It is believed to represent the "late Pleistocene" on the basis of the fauna. At this site, Blanding's turtle, *Emydoidea blandingii*, is south of its presumed modern range in Missouri today.

References. Preston and McCoy (1971), Preston (1979), Hawksley (1986).

11. Zoo Cave, Taney County, Missouri

This vertebrate fauna is derived from a natural cave accumulation that probably involved the activity of predators. It is believed to represent a late Wisconsinan–Holocene transitional sequence.

References. Holman (1974), Hood and Hawksley (1975).

Summary of Herpetological Taxa*

Cryptobranchidae
- *Cryptobranchus* sp.: 11

Proteidae
- *Necturus* cf. *Necturus maculosus:* 5

Ambystomatidae
- *Ambystoma maculatum:* 4, 5
- *Ambystoma opacum:* 2
- *Ambystoma tigrinum:* 4
- *Ambystoma* cf. *Ambystoma tigrinum:* 5
- *Ambystoma* sp.: 1, 5

Plethodontidae
- *Eurycea* sp.: 4, 5
- *Plethodon* cf. *Plethodon dorsalis:* 5
- *Plethodon* cf. *Plethodon glutinosus* complex: 5

Pelobatidae
- *Scaphiopus* cf. *Scaphiopus holbrookii:* 5

Bufonidae
- *Bufo americanus americanus:* 5
- *Bufo woodhousii fowleri:* 1, 11
- *Bufo* sp.: 2, 3, 4, 11

*Numerals in summary correspond to sites described in this section.

Ranidae
Rana catesbeiana: 2, 3, 5
Rana cf. *Rana palustris:* 5
Rana pipiens complex: 5, 11
Rana sp.: 2, 3, 4, 5
Chelydridae
Chelydra serpentina: 3, 10
Emydidae
Emydoidea blandingii: 3, 10
Terrapene carolina ssp. complex: 6
Terrapene carolina: 10
Terrapene cf. *Terrapene ornata:* 3, 4
Terrapene sp.: 3, 7
Trachemys scripta: 6, 10
Testudinidae
cf. *Geochelone* sp.: 7, 8
Crotaphytidae
Crotaphytus sp.: 5
Phrynosomatidae
Sceloporus undulatus: 4, 11
Sceloporus sp.: 5
Teiidae
Cnemidophorus sexlineatus: 11
Cnemidophorus cf. *Cnemidophorus sexlineatus:* 5
Scincidae
Eumeces cf. *Eumeces fasciatus:* 2
Eumeces sp.: 5
Anguidae
Ophisaurus attenuatus: 5, 6, 9
Ophisaurus ventralis: 4
Lacertilia sp.: 3
Colubridae
Carphophis amoenus: 1, 2
Coluber constrictor: 1
Coluber constrictor or *Mastcophis flagellum:* 4, 11
Diadophis punctatus: 2
Elaphe guttata: 1
Elaphe vulpina: 1, 4, 11
Elaphe sp.: 1, 4, 11
Heterodon platirhinos: 4
Heterodon sp.: 1
Lampropeltis (?) *calligaster:* 1
Lampropeltis getula: 1, 4
Lampropeltis triangulum: 1, 2, 4
Lampropeltis sp.: 4

Masticophis flagellum: 1
Nerodia sipedon: 11
Pituophis melanoleucus: 1, 4
Storeria sp.: 2
Thamnophis proximus: 4
Thamnophis sirtalis: 1, 4
Thamnophis proximus or *sirtalis:* 2
Thamnophis sp.: 2
Colubridae sp.: 3
Viperidae
Crotalus horridus: 1, 4, 11
Crotalus sp.: 11
Viperidae sp.: 3
Serpentes sp.: 3
Alligatoridae
(?) *Alligator mississippiensis:* 9

Northern Plains Region

The Northern Plains region has mainly yielded Pleistocene amphibians and reptiles from low-energy aquatic sites such as ponds, oxbows, in pools of braided streams, and off point bars in higher energy streams. Many of the Pleistocene sites of this region were deposited too early in the Pleistocene for radiocarbon dating, thus some early age estimates have been subject to change as additional collections and interpretations have been made. The following northern plains states have yielded important Pleistocene amphibians and reptiles (Fig. 29).

South Dakota
Nebraska
Kansas

Sites of Special Interest

1. Java Local Fauna, South Dakota

The Java local fauna of Walworth County in northern central South Dakota has yielded the northernmost Pleistocene herpetofauna in the Great Plains. The bones accumulated in a low-energy stream situation in sediments of crossbedded fine to medium sand with clay and pebbles included. The fauna is believed to be in the early Irvingtonian Land Mammal Age (IrI; see correction in Repenning, 1987, p. 264). All the herpetological taxa are living today, but several forms now occur south and southeast of the area (*Rana catesbeiana*, *Eumeces septentrionalis*, *Nerodia sipedon*, *Elaphe vulpina*), and one form, *Phrynosoma douglassii*, is presently found about 130 km west of the site.

Reference. Holman (1977b).

Herpetofauna. *Ambystoma tigrinum*, *Bufo woodhousii*, *Rana catesbeiana*, *Rana pipiens* complex, *Phrynosoma douglassii*, *Eumeces septentrionalis*, *Elaphe vulpina*, *Heterodon* sp., *Nerodia sipedon*, *Thamnophis sirtalis*.

2. Albert Ahrens Local Fauna, Nebraska

The Albert Ahrens fauna of Nuckolls County in southeastern Nebraska is the only significant Irvingtonian Pleistocene herpetofauna in the state. It has been suggested that it represents the Kansan glacial age, perhaps the time of the maximum glacial advance (Ford, 1992). The bones were deposited in a low-energy stream environment. All the amphibian and reptile taxa are living today, but some extralimital northern and northeastern (*Ambystoma laterale*, *Rana sylvatica*, *Elaphe vulpina*), eastern (*Bufo americanus*, *Hyla chrysoscelis* or *versicolor*, *Rana clamitans*, *Thamnophis proximus*, *Crotalus horridus*), southern and southeastern (*Pseudacris* cf. *Pseudacris clarkii*, *Regina grahamii*, *Agkistrodon* sp.), and western (*Crotalus viridis*) species indicate there were some range adjustments to the effects of glaciation, and herpetological species adjusted to climatic changes in their own individual ways.

Reference. Ford (1992).

Herpetofauna. *Ambystoma* cf. *Ambystoma laterale*, *Ambystoma tigrinum*, *Ambystoma* sp., *Scaphiopus bombifrons*, *Bufo americanus*, *Bufo cognatus*, *Bufo woodhousii woodhousii*, *Bufo* sp., *Hyla chrysoscelis* or *versicolor*, *Pseudacris* cf. *Pseudacris clarkii*, *Rana catesbeiana*, *Rana clamitans*, *Rana* cf. *Rana blairi*, *Rana sylvatica*, *Rana* sp., *Chelydra serpentina*, *Chrysemys picta* cf. *belli*, *Eumeces obsoletus*, cf. *Coluber constrictor*, *Elaphe obsoleta* or *guttata*, *Elaphe vulpina*, *Elaphe* sp., *Heterodon nasicus*, *Heterodon platirhinos*, *Heterodon* sp., *Lampropeltis triangulum*, *Nerodia sipedon*, *Regina grahamii*, *Storeria* sp., *Thamnophis proximus*, *Thamnophis radix*, *Thamnophis sirtalis*, *Thamnophis* sp., *Agkistrodon* sp., *Crotalus horridus*, *Crotalus viridis*, and *Sistrurus catenatus*.

3. Williams Local Fauna, Kansas

The Williams local fauna of Rice County in the center of Kansas is a significant assemblage of amphibians and reptiles. The vertebrate fossils accumulated in a low-energy aquatic situation in a permanent body of water such as a pond or a slow-moving stream. The fauna is believed to represent the Illinoian glacial age on the basis of the entire faunal assemblage (Holman, 1984). All the herpetological taxa identified are living today, but three species (*Emydoidea blandingii*, *Elaphe vulpina*, *Storeria* cf. *Storeria dekayi*) live northeast of the area.

References. Preston (1979), Holman (1984).

Herpetofauna. *Ambystoma tigrinum*, *Bufo woodhousii woodhousii*, *Acris crepitans*, *Rana catesbeiana*, *Rana pipiens* complex, *Chelydra serpentina*, *Chrysemys picta*, *Emydoidea blandingii*, *Trachemys scripta*, *Coluber constrictor*, *Elaphe vulpina*, *Heterodon* cf. *Heterodon platirhinos*, *Lampropeltis triangulum*, *Nerodia sipedon*, *Pituophis melanoleucus*, *Regina grahamii*, *Sonora semian-*

nulata, *Storeria* cf. *Storeria dekayi*, *Tantilla* sp., *Thamnophis proximus* or *sirtalis*, *Thamnophis radix*, and *Tropidoclonion lineatum*.

4. Sandahl Local Fauna, Kansas

The Sandahl fauna of McPherson County in the center of Kansas is a diverse herpetofauna. It is thought to have been deposited in a low-energy aquatic environment, such as a slow-moving, permanent stream. It is believed to represent the Illinoian glacial age based on the entire faunal assemblage (Holman, 1971). All the herpetological species live in or near the area today.

References. Holman (1971), Preston (1979).

Herpetofauna. *Ambystoma tigrinum*, *Scaphiopus bombifrons*, *Bufo cognatus*, *Bufo woodhousii woodhousii*, *Bufo* sp., *Pseudacris triseriata*, *Rana pipiens* complex, *Trachemys scripta*, *Apalone* sp., *Holbrookia* or *Sceloporus*, *Cnemidophorus* cf. *Cnemidophorus sexlineatus*, *Eumeces* sp., *Coluber* or *Masticophis*, *Heterodon* cf. *Heterodon platirhinos*, *Nerodia sipedon*, *Thamnophis* sp., *Tropidoclonion lineatum*, *Crotalus* cf. *Crotalus viridis*.

5. Kanapolis Local Fauna, Kansas

This local fauna that occurs in Ellsworth County in the center of Kansas is especially noteworthy because of its large number of turtles. The site is the result of a fluvial accumulation both in a small stream and a low-to-base gradient river. The age of site was orginally thought to be Yarmouthian (Holman, 1972) but has been reinterpreted as being Rancholabrean I (Repenning, 1987). A worn fragment of shell may represent the extinct tortoise *Geochelone*, but the remainder of the herpetological taxa are living today. Eastern extralimital herpetological forms (*Hyla chrysoscelis* or *versicolor*, *Sternotherus odoratus*, *Graptemys geographica*, *Eumeces* near *Eumeces fasciatus*), however, occur in the Kanapolis fauna.

References. Holman (1972), Preston (1979).

Herpetofauna. *Acris crepitans*, *Hyla chrysoscelis* or *versicolor*, *Rana catesbeiana*, *Rana pipiens* complex, *Sternotherus odoratus*, *Chelydra serpentina*, *Graptemys geographica*, *Pseudemys* cf. *Pseudemys concinna, Trachemys scripta*, (?) *Geochelone* sp., *Apalone spinifera*, *Apalone* sp., *Eumeces* near *Eumeces fasciatus*, *Ophisaurus attenuatus*, *Coluber* or *Masticophis*, *Heterodon* sp., *Nerodia sipedon*, *Pituophis melanoleucus sayi*, *Regina grahamii*, *Thamnophis proximus* or *sirtalis*, *Sistrurus catenatus*.

6. Mt. Scott Local Fauna, Kansas

The Mt. Scott local fauna of Meade County in southwestern Kansas has a large herpetofauna with a very diverse turtle fauna. It was deposited in an aquatic environment under conditions ranging from moderate-current fluvial to pond depositional. It is considered to be of Illinoian age (e.g., Preston, 1979, Holman, 1987a). The paratype (Fig. 36B) of the extinct turtle *Pseudemys hibbardi* came from this fauna (Preston, 1979). The other herpetological taxa reported are living today, but there were northern (*Emydoidea blandingii*,

Elaphe vulpina) and eastern (*Sternotherus odoratus*, *Terrapene carolina* ssp., *Ophisaurus attenuatus*, *Regina grahamii*) extralimital herpetological species present.

References. Preston (1979), Holman (1987a).

Herpetofauna. *Bufo* sp., (?) *Acris* sp., *Rana pipiens* complex, *Sternotherus odoratus*, *Chelydra serpentina*, *Chrysemys picta*, *Emydoidea blandingii*, *Pseudemys hibbardi* (**paratype locality**), *Terrapene carolina* ssp., *Trachemys scripta*, *Apalone* sp., *Ophisaurus attenuatus*, *Coluber* cf. *Coluber constrictor*, *Diadophis punctatus*, *Elaphe vulpina*, *Heterodon* sp., *Lampropeltis getula*, *Nerodia sipedon*, *Pituophis melanoleucus*, *Regina grahamii*, *Storeria* cf. *Storeria dekayi*, *Thamnophis proximus*, *Thamnophis radix*, and *Sistrurus* or *Crotalus*.

7. Butler Spring Local Fauna, Kansas

The Butler Spring fauna of Meade County in southwestern Kansas is a classic Pleistocene fauna that contains a large assemblage of Pleistocene amphibians and reptiles. The fossil remains accumulated in a low-energy aquatic environment and there appears to have been a "wet" and a "dry" component to the fauna. The Butler Spring fauna is considered to be of Illinoian age (Holman, 1986c). The "sloth locality" of Preston (1979) is here considered to be part of the Butler Spring local fauna. All of the fossil herpetofauna represent taxa that are living today, but there are several northern (*Emydoidea blandingii*, *Elaphe vulpina*) and eastern (*Pseudemys* cf. *Pseudemys concinna heiroglyphica*, *Ophisaurus* cf. *Ophisaurus attenuatus*, *Regina grahamii*, *Storeria* cf. *Storeria dekayi*) extralimital species.

References. Tihen (1962), Preston (1979), Holman (1986c).

Herpetofauna. *Ambystoma tigrinum*, *Bufo cognatus*, *Acris crepitans*, *Rana* sp., *Kinosternon flavescens*, *Chelydra serpentina*, *Chrysemys picta*, *Emydoidea blandingii*, *Pseudemys* cf. *Pseudemys concinna heiroglyphica*, *Terrapene ornata*, *Trachemys scripta*, *Apalone* sp., *Ophisaurus* cf. *Ophisaurus attenuatus*, *Coluber constrictor*, *Diadophis punctatus*, *Elaphe vulpina*, *Heterodon* sp., *Nerodia sipedon*, *Nerodia* sp., *Pituophis melanoleucus*, *Regina grahamii*, *Storeria* cf. *Storeria dekayi*, *Thamnophis proximus*, *Thamnophis radix*, and *Thamnophis* sp.

8. Cragin Quarry Local Fauna, Kansas

This is another classic local fauna in Meade County in southwestern Kansas. This paleobiota is believed to have accumulated in a pond in a semiarid environment. The site is considered to represent the late Sangamonian (Preston, 1979). The Cragin Quarry is the type locality of the medium-sized to large extinct tortoise species *Geochelone equicomes* (Hay, 1917). The type material is under study by Robert Preston.

The Cragin Quarry is the type locality of the extinct toad *Bufo repentinus* Tihen, 1962, who described it on the basis of an ilium with a very high ilial prominence (Fig. 34). The other herpetological taxa represent extant species, but some of them live south (*Cophosaurus* cf. *Cophosaurus texanus*, *Phrynosoma modestum*, *Alligator* sp.) or east (*Pseudacris* cf. *Pseudacris triseriata*, *Terrapene*

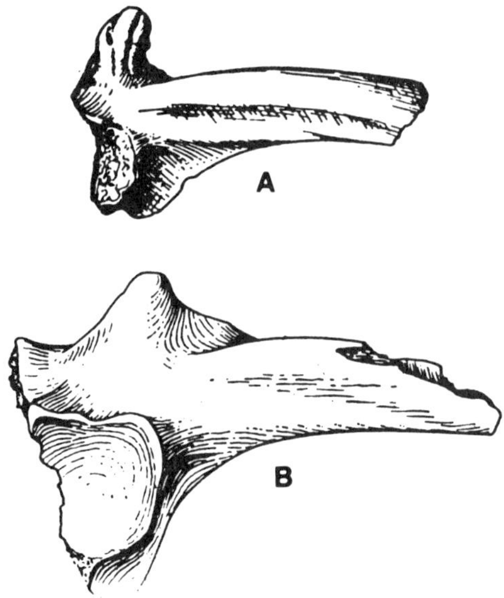

FIGURE 34. (A) Holotype right ilium (University of Michigan Museum of Paleontology Number V34491) of *Bufo repentinus* Tihen from the Rancholabrean of the Cragin Quarry of Meade County, Kansas, compared with (B) a right ilium referred to the extant species *Bufo cognatus* from the same locality. [From Tihen, 1962]

carolina ssp., *Agkistrodon contortrix*) of the area today. The record of *Alligator* sp. is an old one and needs confirmation.

References. Hay (1917), Etheridge (1958, 1960b), Tihen (1960, 1962), Brattstrom (1967), Preston (1979)

Herpetofauna. *Bufo cognatus*, *Bufo repentinus* **(type locality)**, *Pseudacris* cf. *Pseudacris triseriata*, *Kinosternon flavescens*, *Terrapene carolina* ssp., *Geochelone equicomes* **(type locality)**, *Crotaphytus collaris*, *Cophosaurus* cf. *Cophosaurus texanus*, *Phrynosoma cornutum*, *Phrynosoma modestum*, *Cnemidophorus sexlineatus*, *Eumeces obsoletus*, *Coluber constrictor*, (?) *Diadophis*, *Heterodon nasicus*, *Heterodon* near *Heterodon platirhinos*, *Lampropeltis getula*, *Nerodia* sp., *Storeria* sp., *Thamnophis* sp., *Agkistrodon contortrix*, *Crotalus viridis*, and *Alligator* sp.

9. Jinglebob Local Fauna, Kansas

This is the third classic local fauna from Meade County in southwestern Kansas, and it also has a large herpetological component. The fauna accumulated in a low-energy aquatic situation. Based on faunal evidence the Jinglebob herpetofauna is thought to represent the medial Wisconsinan (Preston, 1979). All the herpetological taxa from the Jinglebob fauna are living today, but some of them represent eastern (*Terrapene carolina* ssp., *Ophisaurus attenuatus*, *Elaphe* cf. *obsoleta*) and northern (*Pseudacris* cf. *Pseudacris triseriata*) extralimital forms.

References. Oelrich (1953), Tihen (1954, 1960, 1962), Etheridge (1961), Chantell (1966), Brattstrom (1967), Milstead (1967), Preston (1979).

Herpetofauna. *Ambystoma* sp., *Scaphiopus bombifrons*, (?) *Bufo cognatus*, *Bufo woodhousii woodhousii*, (?) *Acris* sp., *Pseudacris* cf. *Pseudacris triseriata*, *Rana catesbeiana*, *Rana pipiens* complex, *Chelydra serpentina*, *Chrysemys picta*, *Emydoidea blandingii*, *Terrapene carolinia* ssp., *Holbrookia* sp., *Sceloporus* sp., *Eumeces* sp., *Ophisaurus attenuatus*, *Coluber constrictor*, *Elaphe* cf. *Elaphe obsoleta*, *Heterodon nasicus*, *Heterodon platirhinos*, *Lampropeltis calligaster*, *Lampropeltis getula*, *Lampropeltis triangulum*, *Lampropeltis* sp., *Nerodia* sp., *Pituophis* cf. *Pituophis melanoleucus*, and *Thamnophis* sp.

Other Important Sites

10. Kentuck Assemblage, McPherson County, Kansas

This fossil assemblage accumulated in a low-energy aquatic setting. It represents the early Irvingtonian Land Mammal Age (IrI; Repenning, 1987).

References. Tihen (1962), Brattstrom (1967).

11. Arkalon Local Fauna, Seward County, Kansas

This fauna accumulated in a lake, pond, or slow-moving stream. It represents the early Irvingtonian (Irl) based on its supposed occurrence with "unpure" Type O Ash and the presence of the muskrat *Ondatra annectans* (Preston, 1971).

References. Preston (1971, 1979).

12. Courland Canal/Hall Ash Assemblage, Jewell County, Kansas

This fauna is the result of low-energy aquatic sedimentation. It represents the early Irvingtonian Land Mammal Age (IrI; Woodburne, 1987).

Reference. Rogers (1982).

13. Cudahy Fauna, Meade County, Kansas

This assemblage occurred as the result of low-energy sedimentation in an aquatic setting. It represents the Irvingtonian Land Mammal Age (IrII; Woodburne, 1987).

References. Tihen (1960, 1962), Brattstrom (1967).

14. Angus Local Fauna, Nuckolls County, Nebraska

The Angus local fauna was deposited in a low-energy aquatic environment. This faunal assemblage is believed to represent the early Rancholabrean (RlbI; Repenning, 1987) and the Illinoian glacial age (Preston, 1979).

Reference. Preston (1979).

15. Rushville Local Fauna, Sheridan County, Nebraska

The Rushville fauna accumulated in a low-energy aquatic situation. It is believed to represent the Illinoian glacial age (Preston, 1979)

Reference. Preston (1979).

16. Razabeck Local Fauna, Lincoln County, Kansas

This fauna was deposited in a pond or a low gradient stream. It is thought to represent the early Rancholabrean Age (Irl; Repenning, 1987) and the Illinoian glacial age based on the faunal assemblage and a "cooling trend" indicated by some of the mammalian species (Preston, 1979).

References. Brattstrom (1967), Preston (1979).

17. Adams Local Fauna, Meade County, Kansas

The Adams Fauna was deposited in an aquatic situation that produced rather coarse sedimentation. It is considered to represent medial Illinoian (Preston, 1979).

References. Preston and McCoy (1971), Preston (1979).

18. Duck Creek Local Fauna, Ellis County, Kansas

The Duck Creek fauna was deposited in materials that varied from the coarse sands and sandy silts of a stream channel to backwater and oxbow sediments. The deposit is thought to represent the Rancholabrean Land Mammal Age (Woodburne, 1987). Holman (1984) referred to this site as Illinoian.

Reference. Holman (1984).

19. XI Ranch Localities, Meade County Kansas

These localities are considered as one unit by Preston (1979). They all represent aquatic sedimentary processes and are considered to represent the Rancholabrean Land Mammal Age and the early Sangamonian (Preston, 1979).

References. Galbreath (1948), Preston (1979).

20. Jones Fauna, Meade County, Kansas

This fauna was the result of a natural accumulation in a lacustrine situation. A radiocarbon date of 29,000 to 26,700 ybp on shell indicates a medial to late Wisconsinan age (Lundelius et al., 1983).

References. Tihen (1960, 1962), Brattstrom (1967), Preston (1979).

21. Robert Fauna, Meade County, Kansas

The Robert fauna accumulated in a marsh setting. A radiocarbon date of 11,100 $\pm$ 390 ybp on shell indicates a very late Wisconsinan age (Lundelius et al., 1983).

Reference. Holman (1987b).

Summary of Herpetological Taxa*

Ambystomatidae
Ambystoma cf. *Ambystoma laterale:* 2

*Numerals in summary correspond to sites described in this section.

Ambystoma maculatum: 12
Ambystoma tigrinum: 1, 2, 3, 4, 7, 12, 18, 20
Ambystoma sp.: 2, 9
Pelobatidae
Scaphiopus bombifrons: 2, 4, 9
Bufonidae
Bufo americanus: 2
Bufo cognatus: 2, 4, 7, 8, 20
(?) *Bufo cognatus:* 9
Bufo hemiophrys: 12
Bufo cf. *Bufo hemiophrys:* 13
Bufo repentinus: 8
Bufo woodhousii woodhousii: 2, 3, 4, 9, 18, 20
Bufo woodhousii: 1
Bufo sp.: 2, 4, 6, 10
Hylidae
Acris crepitans: 3, 5, 7
(?) *Acris* sp.: 6, 9
Hyla chrysoscelis or *versicolor:* 2, 5
Pseudacris cf. *Pseudacris clarkii:* 2
Pseudacris triseriata: 4
Pseudacris cf. *Pseudacris triseriata:* 8, 9, 13, 20
Ranidae
Rana catesbeiana: 1, 2, 3, 5, 9, 12
Rana cf. *Rana blairi:* 2
Rana clamitans: 2
Rana pipiens complex: 1, 3, 4, 5, 6, 9, 12, 18
Rana sylvatica: 2, 18
Rana sp.: 2, 7
Kinosternidae
Kinosternon flavescens: 7, 8, 16
Kinosternon sp.: 19
Sternotherus odoratus: 5, 6
Chelydridae
Chelydra serpentina: 2, 3, 5, 6, 7, 9, 11, 14, 15, 16, 19
Emydidae
Chrysemys picta cf. *belli:* 2
Chrysemys picta: 3, 6, 7, 9, 11, 12, 14, 15, 16, 19, 20
Emydoidea blandingii: 3, 6, 7, 9, 11, 15, 17, 19
Graptemys geographica: 5
Graptemys geographica/ kohni: 19
Pseudemys cf. *Pseudemys concinna heiroglyphica:* 7
Pseudemys cf. *Pseudemys concinna:* 5
Pseudemys cf. *Pseudemys floridana hoyi:* 11
Pseudemys hibbardi: 6

Terrapene carolina ssp.: 6, 8, 9
Terrapene ornata ornata: 11
Terrapene ornata: 7, 15, 19
Trachemys scripta: 3, 4, 5, 6, 7, 11, 12, 14, 17, 19

Testudinidae
Geochelone equicomes: 8
Geochelone near *Geochelone equicomes:* 14
Geochelone small species: 11
Geochelone sp.: 12
(?) *Geochelone* sp.: 5
Gopherus large extinct species: 11, 19

Trionychidae
Apalone spinifera: 5, 15
Apalone sp.: 4, 5, 6, 7, 11, 19

Crotaphytidae
Crotaphytus collaris: 8

Phrynosomatidae
Cophosaurus cf. *Cophosaurus texanus:* 8
Holbrookia sp.: 9
Sceloporus sp.: 9
Holbrookia or *Sceloporus:* 4
Phrynosoma cornutum: 8
Phrynosoma douglassii: 1
Phrynosoma modestum: 8

Teiidae
Cnemidophorus sexlineatus: 8
Cnemidophorus cf. *Cnemidophorus sexlineatus:* 4

Scincidae
Eumeces near *Eumeces fasciatus:* 5
Eumeces obsoletus: 2, 8
Eumeces septentrionalis: 1
Eumeces sp.: 4, 9, 13

Anguidae
Ophisaurus attenuatus: 5, 6, 9
Ophisaurus cf. *Ophisaurus attenuatus:* 7

Colubridae
Coluber constrictor: 3, 7, 8, 9, 12, 21
Coluber cf. *Coluber constrictor:* 6
cf. *Coluber constrictor:* 2
Coluber or *Masticophis:* 4, 5
Diadophis punctatus: 6, 7, 21
(?) *Diadophis:* 8
Elaphe obsoleta: 10, 20
Elaphe cf. *Elaphe obsolata:* 9
Elaphe obsoleta or *guttata:* 2

Elaphe vulpina: 1, 2, 3, 6, 7
Elaphe sp.: 2
Heterodon nasicus: 2, 8, 9, 12
Heterodon platirhinos: 2, 9, 10
Heterodon cf. *Heterodon platirhinos:* 3, 4
Heterodon near *Heterodon platirhinos:* 8
Heterodon sp.: 1, 2, 5, 6, 7, 21
Lampropeltis calligaster: 9, 18
Lampropeltis getula: 6, 8, 9, 21
Lampropeltis triangulum: 2, 3, 9, 10
Lampropeltis sp.: 9
Nerodia sipedon: 1, 2, 3, 4, 5, 6, 7, 12, 21
Nerodia cf. *Nerodia sipedon:* 16
Nerodia sp.: 7, 8, 9, 10, 13
Pituophis melanoleucus sayi: 5
Pituophis melanoleucus: 3, 6, 7
Pituophis cf. *Pituophis melanoleucus:* 9
Regina grahamii: 2, 3, 5, 6, 7, 21
Sonora semiannulata: 3
Storeria cf. *Storeria dekayi:* 3, 6, 7, 21
Storeria sp.: 2, 8, 13
Tantilla sp.: 3
Thamnophis proximus: 2, 6, 7, 12, 21
Thamnophis radix: 2, 3, 6, 7, 12, 21
Thamnophis proximus or *sirtalis:* 3, 5, 21
Thamnophis sirtalis: 1, 2, 12
Thamnophis sp.: 2, 4, 7, 8, 9, 13, 20
Tropidoclonion lineatum: 3, 4

Viperidae
Agkistrodon contortrix: 8, 12
Agkistrodon sp.: 2
Crotalus horridus: 2
Crotalus viridis: 2, 8
Crotalus cf. *Crotalus viridis:* 4
Sistrurus catenatus: 2, 5
Sistrurus or *Crotalus:* 6

Alligatoridae
Alligator sp.: 8

Southern Plains Region

The Southern Plains region has Pleistocene herpetofaunas derived from both cave and aquatic situations. It is a region that has yielded a large number of herpetofaunas, some of them quite diverse. The following states are included in this section (Fig. 29):

Oklahoma
Texas

Note. Extinct tortoises of the genera *Geochelone* and *Gopherus*, other than the small extinct tortoise *Geochelone wilsoni*, are not listed as to species, but merely as size categories "large" or "small" following Preston (1979). The status of the currently recognized large extinct species *Gopherus laticauda* (see Preston, 1979) will be discussed in Chapter 8.

Sites of Special Interest

1. Gilliland Fauna, Texas

The Gilliland fauna in Knox County in northwest Texas near the panhandle provides a window on early Pleistocene herpetological life in Texas. The fauna accumulated in an environment of medium- to low-energy aquatic sedimentation. The site represents the early Irvingtonian Land Mammal Age (IrI; Repenning, 1987). Both a large and a small species of extinct tortoise (*Geochelone*) are present, and an extinct subspecies of slider turtle, *Trachemys scripta bisornata*, has been reported from the site. All the other herpetological species represent modern taxa and most live in northern Texas today with the exception of the southern extralimital *Gopherus* near *Gopherus polyphemus* and the eastern extralimital *Eumeces* cf. *Eumeces fasciatus*.

References. Preston (1966), Parmley (1988a).

Herpetofauna. *Bufo woodhousii woodhousii*, *Rana pipiens* complex, *Terrapene* sp., *Trachemys scripta bisornata*, *Trachemys* or *Pseudemys*, *Geochelone* large species, *Geochelone* small species, *Gopherus* near *Gopherus polyphemus*, *Apalone spinifera*, *Eumeces* cf. *Eumeces fasciatus*, *Heterodon* sp., *Lampropeltis calligaster* or *getula*, *Nerodia* cf. *Nerodia erythrogaster*, *Rhinocheilus lecontei*, *Thamnophis* sp., *Tropidoclonion lineatum*, *Crotalus* sp., and *Alligator* sp.

2. Fyllan Cave, Texas

This site in northwest Austin, Travis County, Texas, has yielded the only large Irvingtonian cave herpetofauna in the Southern Plains region. The bones are the result of a natural accumulation in a cave and there are strong indications that many of them were transported there from some distance. The site is considered to represent the later Irvingtonian Land Mammal Age (IrII; Repenning, 1987). Aside from a large extinct tortoise (*Geochelone*), all the taxa from the site are living today, but there is a very large glass lizard (*Ophisaurus* sp.) that might represent an extinct form. There are also extralimital eastern (*Notophthalmus* sp.) and southwestern (*Crotaphytus* cf. *Crotaphytus reticulatus*) forms in the fauna.

Reference. Holman and Winkler (1987).

Herpetofauna. *Notophthalmus* sp., *Ambystoma tigrinum*, *Bufo* sp., *Kinosternon flavescens*, *Terrapene carolina* ssp., *Terrapene* sp., *Geochelone* large species, *Crotaphytus* cf. *Crotaphytus reticulatus*, *Sceloporus* near *Sceloporus olivaceous*, *Sceloporus* near *Sceloporus undulatus*, *Eumeces fasciatus*, *Eumeces obsoletus*, *Ophisaurus* "giant form," *Coluber* or *Masticophis*, *Elaphe guttata*, *Lampropeltis getula*, *Lampropeltis triangulum*, *Nerodia fasciata*, *Storeria* sp., *Tantilla* sp., *Thamnophis* sp., *Tropidoclonion lineatum*, *Micrurus fulvius*, *Crotalus atrox*, and *Crotalus horridus*.

3. Slaton Local Fauna, Texas

The Slaton local fauna at the base of the panhandle in Lubbock County, Texas, was deposited in the muck of an ancient lake. This fauna is considered to represent the earlier part of the Rancholabrean Land Mammal Age (RlbI, Woodburne, 1987), but has been interpreted differently sometimes (see Preston, 1979, p. 17). The herpetological members of the Slaton fauna would not be found living in the area today with the exception of the extralimital southern *Alligator mississippiensis* and the extralimital eastern *Terrapene carolina* ssp. No extinct herpetological taxa have been found at this site.

References. Milstead (1967), Holman (1969b, 1969c), Preston (1979).

Herpetofauna. *Ambystoma tigrinum*, *Bufo cognatus* or *speciosus*, *Bufo* cf. *Bufo woodhousii*, *Acris crepitans*, *Pseudacris clarkii*, *Rana pipiens* complex, *Terrapene carolina* ssp., *Crotaphytus collaris*, *Holbrookia* sp., *Phrynosoma cornutum*, *Sceloporus undulatus*, *Coluber* or *Masticophis*, *Lampropeltis getula*, *Nerodia erythrogaster*, *Rhinocheilis lecontei*, *Thamnophis* cf. *Thamnophis proximus*, *Crotalus atrox*, and *Alligator mississippiensis*.

4. Easley Ranch Local Fauna, Texas

This site occurs in Foard County in north-central Texas. It has yielded a quite diverse herpetofauna. The bones were deposited in a low-energy aquatic area such as a marsh or a shallow lake and the fossils came from a clayey matrix. This Rancholabrean site is somewhat transitional between Sangamonian and Wisconsinan times (Preston, 1979). Extinct species include two large tortoises (*Geochelone* and *Gopherus*); an extralimital southern frog (*Syrrhophus marnockii*) and an eastern snake (*Thamnophis sirtalis*) are also present.

References. Holman (1962b), Lynch (1964), Preston (1979).

Herpetofauna. *Ambystoma* sp., *Syrrhophus marnockii*, *Bufo woodhousii*, *Acris* cf. *Acris crepitans*, *Hyla chrysoscelis* or *versicolor*, *Pseudacris clarkii*, *Rana pipiens* complex, *Geochelone* large extinct species, *Gopherus* large extinct species, *Holbrookia* sp., *Coluber* or *Masticophis*, *Heterodon* sp., *Opheodrys aestivus*, *Storeria* cf. *Storeria dekayi*, *Thamnophis* cf. *Thamnophis proximus*, *Thamnophis sirtalis*, and *Thamnophis* sp.

5. Clear Creek Fauna, Texas

The Clear Creek site occurs in Denton county near Fort Worth, Texas. This site yielded a very diverse herpetofauna, including several species of aquatic turtles. The bones were deposited in a permanent but low-energy aquatic situation such as a slough or shallow lake. A radiocarbon date of 28,000 ybp places the fauna in the medial to late Wisconsinan range. The fauna consists of two distinct components, one aquatic and one xerophytic terrestrial, and of species that represent a mingling of eastern and western forms. All the herpetological forms, except for a large extinct tortoise (*Geochelone*), represent living taxa, although somewhat eastern (*Ambystoma texanum*, *Sternotherus odoratus*, *Terrapene carolina* ssp.) and western (*Arizona elegans*) extralimital forms occur in the fauna.

References. Holman (1963), Preston (1979).

Herpetofauna. *Ambystoma texanum*, *Bufo* cf. *Bufo woodhousii*, *Acris* cf. *Acris crepitans*, *Gastrophryne olivacea*, *Rana pipiens* complex, *Sternotherus odoratus*, *Terrapene carolina* ssp., *Terrapene ornata*, *Trachemys scripta*, *Geochelone* large species, *Apalone* sp., *Cnemidophorus* cf. *Cnemidophorus gularis*, *Ophisaurus* sp., *Arizona elegans*, *Coluber* or *Masticophis*, *Heterodon nasicus*, *Lampropeltis triangulum*, *Nerodia erythrogaster* or *sipedon*, *Opheodrys aestivus*, *Thamnophis* cf. *Thamnophis proximus*, *Virginia striatula*, *Agkistrodon piscivorous*, and *Crotalus* sp.

6. Howard Ranch Local Fauna, Texas

The Howard Ranch fauna (sometimes called the Groesbeck Creek site, e.g., Holman, 1964) occurs in Hardeman County at the eastern base of the Texas panhandle. This rich herpetological site accumulated in a basin fill that was probably a broad marshy area at the time. The fauna from this basin was actually taken from two localities with radiocarbon dates ca. 17,000 and 19,000 ybp (Preston, 1979), thus the late Wisconsinan is represented. All the herpetological taxa of this fauna are extant today with the exception of the questionable treefrog species *Hyla holmani* Lynch, 1966, which was described on the basis of an ilium (Fig. 35). This species will be discussed again in Chapter 8. Most of the extant herpetological taxa could be found living in the area today with the exception of the somewhat eastern forms *Ambystoma texanum*, *Hyla versicolor*, *Pseudacris streckeri*, *Terrapene carolina* ssp., and *Agkistrodon contortrix*.

References. Holman (1964), Preston (1979).

Herpetofauna. *Ambystoma texanum*, *Bufo cognatus* or *speciosus*, *Acris crepitans*, *Hyla chrysoscelis* or *versicolor*, *Hyla holmani* **(type locality)**, *Pseudacris streckeri*, *Rana pipiens* complex, *Chelydra serpentina*, *Terrapene carolina* ssp., *Trachemys* or *Pseudemys*, *Crotaphytus collaris*, *Arizona elegans*, *Coluber* or *Masticophis*, *Heterodon nasicus*, *Lampropeltis getula*, *Lampropeltis triangulum*, *Nerodia erythrogaster* or *sipedon*, *Thamnophis marcianus* or *radix*, *Thamnophis proximus*, *Agkistrodon contortrix*, and *Crotalus* sp.

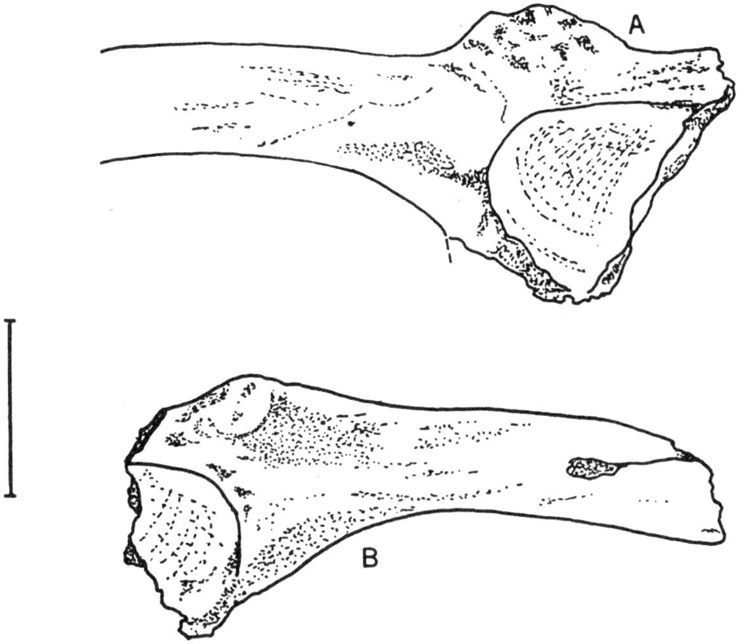

FIGURE 35. (A) Holotype left ilium (Midwestern University Number 6581) of *Hyla holmani* Lynch and (B) referred right ilium of this species from the Rancholabrean Howard Ranch fauna of Hardeman County, Texas. The line equals 1 mm and refers to both A and B. [From Lynch, 1966]

7. Schulze Cave Fauna, Texas

The Schulze Cave fauna comes from Edwards County on the Edwards plateau west of Austin, Texas. It has been suggested that the fauna was brought in by barn owls (*Tyto alba*), entered voluntarily, or fell into the manholelike opening. Bones came from layers radiocarbon dated at about 9,000 ybp and deeper layers estimated at representing an age of from about 12,000 to 13,000 ybp. The site thus represents very late Wisconsinan times. Except for a small extinct tortoise (*Geochelone*), all the herpetological taxa are extant and all but one snake, *Elaphe obsoleta*, which occurs east of the site, would be expected to live near the cave today. *Bufo woodhousii bexarensis,* listed by Holman (1969a) from the site, has been reassigned to *Bufo woodhousii woodhousii* in the list below based on the discussion in Parmley (1986).

References. Holman (1969a), Parmley (1986).

Herpetofauna. *Ambystoma tigrinum*, *Scaphiopus* cf. *Scaphiopus couchii*, *Hylactophryne augusti latrans*, *Syrrhophus marnockii*, *Bufo woodhousii woodhousii*, *Hyla chrysoscelis* or *versicolor*, *Terrapene ornata*, *Geochelone* cf. *Geochelone wilsoni*, *Crotaphytus collaris*, *Sceloporus undulatus*, *Eumeces* cf. *Eumeces obsoletus*, *Arizona elegans*, *Coluber* or *Masticophis*, *Elaphe* cf. *Elaphe guttata*, *Elaphe obsoleta*, *Heterodon* cf. *Heterodon nasicus*, *Heterodon* cf. *Heterodon platirhinos*, *Lampropeltis getula*, *Lampropeltis triangulum*, *Opheo-*

drys cf. *Opheodrys aestivus*, *Pituophis melanoleucus*, *Thamnophis* sp., *Agkistrodon* sp., and *Crotalus* sp.

8. Lubbock Lake Site, Texas

The Lubbock Lake site in Lubbock County, in the staked-plains or Llano Estacado area of western Texas, is an important locality because of the association of herpetological species with Paleo-Indian, Archaic, Ceramic, Protohistoric, and historic human deposits (Johnson, 1987). The Clovis period (11,100 ybp), the Folsom period (10,800 to 10,300 ybp), and the Plainview period (10,000 ypb) are the oldest horizons at the Lubbock Lake site. Thomas Van Devender identified the herpetological material in the definitive Johnson (1987) report on the vertebrates of the site.

Of considerable interest is that the Clovis point horizon marks the last occurrence at the Lubbock Lake site of the extinct toad, *Bufo woodhousii bexarensis,* the large extinct box turtle, *Terrapene carolina* spp., the small extinct tortoise, *Geochelone wilsoni,* and the large extinct tortoise, *Geochelone* sp. On the other hand, the modern box turtle, *Terrapene ornata* does not show up until the Plainview period and is common thereafter (Johnson, 1987). Extralimital eastern species from the Clovis horizon include *Rana palustris* and *Virginia* cf. *Virginia striatula.*

Reference. Johnson (1984, 1987)

Herpetofauna. Clovis period: *Ambystoma tigrinum, Scaphiopus bombifrons* or *hammondii, Bufo cognatus, Bufo woodhousii bexarensis, Bufo* sp., *Acris crepitans, Rana catesbeiana, Rana palustris, Rana pipiens* complex, *Kinosternon flavescens, Chelydra serpentina, Terrapene carolina* spp., *Trachemys Scripta, Geochelone wilsoni, Geochelone* large extinct species, *Apalone* sp., *Carphophis amoenus, Elaphe guttata, Gyalopion canum, Heterodon nasicus, Lampropeltis triangulum, Nerodia* sp., *Salvadora* sp., *Sonora semiannulata, Thamnophis* cf. *Thamnophis sirtalis, Tropidoclonion lineatum, Virginia* cf. *Virginia striatula,* and *Crotalus atrox.* Folsom period: *Rana catesbeiana, Rana pipiens* complex, *Kinosternon flavescens, Trachemys scripta, Elaphe* or *Pituophis,* and *Thamnophis* cf. *Thamnophis sirtalis.* Plainview period: *Ambystoma tigrinum, Bufo* sp., *Rana catesbeiana, Rana pipiens* complex, *Kinosternon flavescens, Chelydra serpentina, Terrapene ornata, Trachemys scripta, Eumeces obsoletus, Coluber constrictor, Lampropeltis getula, Nerodia cyclopion* or *rhombifera, Sonora semiannulata, Thamnophis marcianus, Thamnophis proximus, Thamnophis* cf. *Thamnophis sirtalis, Tropidoclonion lineatum,* and *Agkistrodon contortrix.*

9. Cave Without a Name, Texas

The Cave Without a Name fauna of Kendall County, just north of Austin in southern Texas on the Edwards plateau, is significant because of the very large herpetofauna that it contains. The fossils come from a red clay horizon and are thought to have mainly been derived from pellets cast by owls. A radiocarbon date of 10,980 $\pm$ 190 ybp on bone places this fauna very late in the Wisconsinan. All the herpetological taxa reported from this cave are living today, but some intrusive eastern (*Terrapene carolina* spp., *Lampropeltis calligaster*, *Thamnophis sirtalis*, *Agkistrodon piscivorous*) and southern (*Sceloporus variabilis*, *Eumeces tetragrammus*) forms are present.

References. Holman (1968), Hill (1971).

Herpetofauna. *Plethodon albagula*, cf. Leptodactylidae sp., cf. *Rana* sp., *Terrapene carolina* ssp., *Crotaphytus collaris*, *Cophosaurus texanus*, *Phrynosoma cornutum*, *Sceloporus poinsettii*, *Sceloporus undulatus*, *Sceloporus variabilis*, *Cnemidophorus* sp., *Eumeces obsoletus*, *Eumeces tetragrammus*, *Arizona elegans*, *Coluber* or *Masticophis*, *Coluber constrictor*, *Diadophis punctatus*, *Elaphe guttata*, *Elaphe* sp., *Heterodon nasicus*, *Heterodon platirhinos*, *Lampropeltis calligaster*, *Lampropeltis getula*, *Lampropeltis triangulum*, *Lampropeltis* sp., *Masticophis* sp., *Nerodia erythrogaster* or *fasciata*, *Opheodrys* sp., *Pituophis melanoleucus*, *Rhinocheilus lecontei*, *Tantilla* sp., *Thamnophis proximus*, *Thamnophis sirtalis*, *Thamnophis* sp., *Micrurus fulvius*, *Agkistrodon contortrix*, *Agkistrodon piscivorous*, *Agkistrodon* sp., and *Crotalus* sp.

10. Miller's Cave Fauna, Texas

This Miller's Cave site of Llano County, north of Austin in the Edwards plateau region of Texas, contains a large herpetofauna. The bones are the result of a natural accumulation in the cave. A radiocarbon date on material from the cave was too young because the dating technique of the cave fauna was not good (Holman, 1966b). Nevertheless, a very late Wisconsinan age for the cave fauna is indicated as the only extinct mammalian species is the armadillo *Dasypus bellus*. None of the herpetological taxa are extinct and all occur in the vicinity of the fossil site today.

References. Patton (1963), Holman (1966b).

Herpetofauna. *Ambystoma tigrinum*, *Pseudacris streckeri*, *Rana catesbeiana*, *Rana pipiens* complex, *Rana* sp., *Crotaphytus collaris*, *Crotaphytus* sp., *Sceloporus* sp., *Cnemidophorus sexlineatus*, *Cnemidophorus* sp., *Eumeces obsoletus*, *Arizona elegans*, *Coluber* or *Masticophis*, *Elaphe* or *Pituophis*, *Elaphe* sp., *Lampropeltis calligaster*, *Lampropeltis triangulum*, *Opheodrys aestivus*, *Opheodrys* sp., *Tantilla* sp., *Thamnophis proximus*, *Crotalus atrox*, and *Crotalus* sp.

11. Fowlkes Cave, Texas

This cave lies in Culberson County in trans-Pecos Texas. It is the only substantial Pleistocene herpetofauna from this part of the state. Most of the bones are believed to have been derived by the predatory activity of owls. The age of the fossil-bearing layer is estimated to be late Pleistocene based on the presence of the extinct antilocaprid (*Capromeryx*) and the long-nosed peccary (*Mylophys*). None of the herpetological taxa are extinct, but extralimital southeastern species (*Bufo valliceps* and *Opheodrys aestivus*) occur in the cave fauna.

References. Parmley (1988b, 1990).

Herpetofauna. *Bufo valliceps*, *Bufo woodhousii*, *Rana pipiens* complex,

Arizona elegans, *Bogartophis subocularis*, *Coluber* or *Masticophis*, *Diadophis* cf. *Diadophis punctatus*, *Elaphe* cf. *Elaphe guttata*, *Gyalopion canum*, *Hypsiglena torquata*, *Lampropeltis getula*, *Lampropeltis mexicana*, *Lampropeltis triangulum*, *Nerodia erythrogaster*, *Opheodrys aestivus*, *Pituophis melanoleucus*, *Rhinocheilus lecontei*, *Sonora* sp., *Tantilla* sp., *Thamnophis* cf. *Thamnophis marcianus*, *Thamnophis* sp., and *Crotalus* sp.

Other Important Sites

12. Vera Local Fauna, Knox County, Texas

The Vera fauna was deposited mainly under low-energy aquatic conditions. It is considered to represent the Irvingtonian Land Mammal Age (IrII; Repenning, 1987).

Reference. Parmley (1988a).

13. Berends Local Fauna, Beaver County, Oklahoma

The Berends fauna accumulated mainly in a low-energy aquatic situation. It is believed to represent the Illinoian age.

Reference. Preston (1979).

14. Doby Springs Local Fauna, Harper County, Oklahoma

The Doby Springs fauna accumulated in a low-energy, permanent aquatic situation. It is thought to represent the Illinoian age. The Doby Springs local fauna yielded the holotype plastral lobe (Fig. 36) of the extinct emydid turtle *Pseudemys hibbardi* (Preston, 1979). This species is questionable and its status will be further discussed in Chapter 8.

Reference. Preston (1979).

15. Nye Sink Locality, Beaver County, Oklahoma

This sink deposit is considered to be of Rancholabrean age and to represent the early part of the Sangamonian (Preston, 1979). It is of interest because the extralimital northern *Emydoidea blandingii* and a large extinct tortoise, *Gopherus* sp., occur together in the fauna.

References. Preston (1979).

16. Pitt Bridge Localities, Brazos and Burleson Counties, Texas

These localities developed as shallow ponds and are believed to represent the Sangamonian age.

References. Holman (1969b), Milstead (1967), Auffenberg (1974), Preston (1979).

17. Lewisville Site, Denton County, Texas

This low-energy fluvial site probably represents late medial Wisconsinan times (Preston, 1979). It is thought to be of about the same age as the Moore Pit fauna

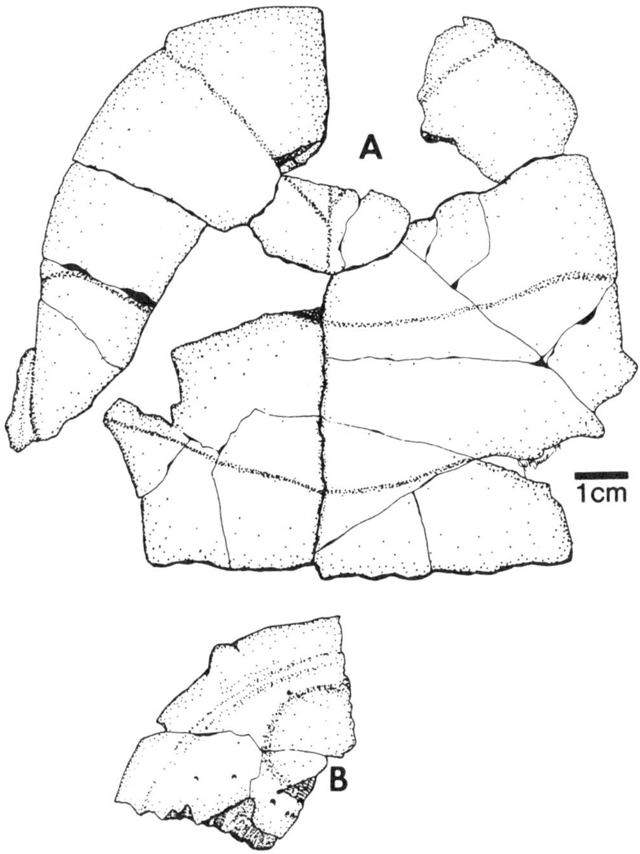

FIGURE 36. (A) Holotype plastral forelobe (University of Michigan Museum of Paleontology Number V33333) of *Pseudemys hibbardi* Preston from the Rancholabrean Doby Springs local fauna of Harper County, Oklahoma; and (B) left epiplastron (UMMP V61846) of the same species from the Rancholabrean Mount Scott Local fauna of Meade County, Kansas. The line equals 1 cm and applies to both A and B. [From Preston, 1979]

(site 18). A huge box turtle (*Terrapene carolina* ssp., Fig. 42E) came from the Lewisville site.

References. Holman (1965c), Milstead (1967), Preston (1979).

18. Moore Pit Local Fauna, Dallas County, Texas

The Moore Pit local fauna came from river terrace deposits near Dallas, Texas. It probably represents late medial Wisconsinan times (Preston, 1979). The complete shell of a very large *Geochelone*, figured in the classic paper of Hibbard (1960), came from this site. Alligator remains also occurred there.

References. Milstead (1967), Holman (1969b), Preston (1979).

19. Carrol Creek Local Fauna, Donley County, Texas

This local fauna was found in a sinkhole deposit in the lower Texas panhandle. A radiocarbon date of 32,400 ± 560 ybp on mollusk shells indicates that the site was deposited during late medial Wisconsinan times.

Reference. Kasper and Parmley (1990).

20. Sims Bayou Fauna, Harris County, Texas

This low-energy fluvial deposit from the south bank of Sims Bayou in Houston, Texas, probably represents late medial Wisconsinan times (Holman, 1965d). Both large *Geochelone* and *Alligator* remains were found in this deposit.

References. Holman (1965d), Preston (1979).

21. Batte Gravel Pit, Milam County, Texas

This fluvial deposit is considered to represent the late Pleistocene on the basis of the presence of *Terrapene carolina* ssp. (Preston, 1979).

Reference. Preston (1979).

22. Ingleside Fauna, San Patricio County, Texas

This fauna from coastal south Texas near Corpus Christi was derived from calcareous pond sands and marls overlying a marine lagoonal clay. The site is thought to represent post Sangamonian late Pleistocene times.

Reference. Preston (1979).

23. Berclair Local Fauna, Bee County, Texas

This fauna occurs in localities near Beeville in southern Texas. The turtles indicate that the fauna may have accumulated in permanent ponds or some other permanent low-energy aquatic situations. The site is considered to be late Pleistocene.

References. Milstead (1967), Holman (1969b), Preston (1979).

24. Friesenhahn Cave, Bexar County, Texas

The Friesenhahn Cave north of San Antonio in the Edwards plateau is a deposit where the majority of bones are thought to have been derived by the activities of owls and mammalian carnivores. Several layers have been sampled over the years. Radiocarbon dates range from 19,000 to ca. 10,000 ybp, so that a late Wisconsinan age is indicated. The herpetofauna listed from this site is small, but it is the type locality of two extinct taxa.

An extinct subspecies of toad, *Bufo woodhousii bexarensis* Mecham, 1959, is based on a large femur (holotype) and other elements (Fig. 37). This form will be addressed again in Chapter 8. The holotype shell (Fig. 43C) of a small extinct tortoise, *Geochelone wilsoni* (Milstead, 1956) also came from this cave. This extinct species will also be addressed in Chapter 8.

References. Milstead (1956, 1967), Auffenberg (1958, 1974), Mecham (1959), Holman (1969b), Preston (1979).

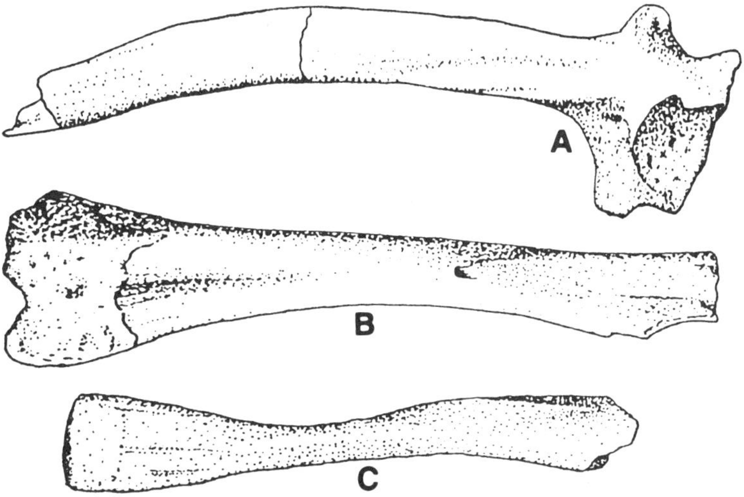

FIGURE 37. *Bufo woodhousei bexarensis* Mecham from the Rancholabrean Friesenhahn Cave locality, Bexar County, Texas. (A) Left ilium; (B) right tibiofibula; (C) holotype left femur (Texas Memorial Museum Number 933-3708A). [From Mecham, 1959]

25. Ben Franklin Local Fauna, Fannin and Delta Counties, Texas

The bones of this site represent a natural accumulation in a fluvial setting. The site represents very latest Wisconsinan or very earliest Holocene times as radiocarbon dates of 11,135 ± 450 and 9,550 ± 377 ybp have been yielded by mollusk shell.

Reference. Holman (1963).

26. Domebo Locality, Caddo County, Oklahoma

This fauna was found in context with early humans of the Plano Culture which existed ca. 11,000 ybp, suggesting a very late Wisconsinan age. It is an important site as the small extinct tortoise *Geochelone wilsoni* was found there.

References. Milstead (1967), Preston (1979).

Summary of Herpetological Taxa*

Salamandridae
Notophthalmus sp.: 2
Amphiumidae
Amphiuma means: 20
Ambystomatidae
Ambystoma texanum: 5, 6, 20

*Numerals in summary correspond to sites described in this section.

Ambystoma tigrinum: 2, 3, 7, 8, 10, 12, 19, 25
Ambystoma sp.: 4
Plethodontidae
Plethodon albagula: 9
Pelobatidae
Scaphiopus bombifrons or *hammondii:* 8
Scaphiopus cf. *Scaphiopus couchii:* 7
Scaphiopus couchii or *hammondii:* 24
Leptodactylidae
Hylactophryne augusti latrans: 7
Hylactophryne augusti: 24
Syrrhophus marnockii: 4, 7
Syrrhophus cf. *Syrrhophus marnockii:* 12
cf. Leptodactylidae sp.: 9
Bufonidae
Bufo cognatus: 8
Bufo (?) *cognatus:* 24
Bufo cognatus or *speciosus:* 3, 6
Bufo cf. *Bufo hemiophrys:* 12
Bufo valliceps: 11
Bufo woodhousii bexarensis: 8, 24
Bufo woodhousii woodhousii: 1, 7
Bufo woodhousii: 4, 11
Bufo cf. *Bufo woodhousii:* 3, 5
Bufo sp.: 2, 8, 12, 14
Hylidae
Acris crepitans: 3, 6
Acris cf. *Acris crepitans:* 4, 5, 12
Hyla chrysoscelis or *versicolor:* 4, 6, 7
Hyla holmani: 6
Pseudacris clarkii: 3, 4
Pseudacris streckeri: 6, 10
Pseudacris cf. *Pseudacris triseriata:* 14
Microhylidae
Gastrophryne olivacea: 5
Ranidae
Rana catesbeiana: 8, 10, 22
Rana cf. *Rana catesbeiana:* 25
Rana pipiens complex: 1, 3, 4, 5, 6, 8, 10, 11, 20, 24
Rana sp.: 10, 12, 19, 24
cf. *Rana* sp.: 9
Kinosternidae
Kinosternon flavescens: 2, 8, 13, 23
Sternotherus odoratus: 5, 13

Chelydridae
Chelydra serpentina: 6, 8, 13, 14, 15
Emydidae
Chrysemys picta: 13, 14
Emydoidea blandingii: 13, 14, 15
Graptemys geographica or *kohni:* 16, 21
Graptemys sp.: 18
(?) *Pseudemys concinna:* 23
Pseudemys hibbardi: 14
Terrapene carolina ssp.: 2, 3, 5, 6, 8, 9, 16, 17, 18, 20, 21, 22, 23, 24, 26
Terrapene ornata: 5, 7, 8, 13, 14
Terrapene sp.: 1, 2
Trachemys scripta bisornata: 1
Trachemys scripta: 5, 8, 13, 14, 20, 21, 22, 23
Trachemys or *Pseudemys:* 1, 6
Testudinidae
Geochelone wilsoni: 8, 24, 26
Geochelone cf. *Geochelone wilsoni:* 7
Geochelone large extinct species: 1, 2, 4, 5, 16, 17, 18, 20, 22
Geochelone small extinct species: 13
Gopherus large extinct species: 4, 15, 22
Gopherus near *Gopherus polyphemus:* 1
Trionychidae
Apalone spinifera: 1, 16, 22
Apalone sp.: 5, 8, 13, 18, 20
Crotaphytidae
Crotaphytus collaris: 3, 6, 7, 9, 10, 24
Crotaphytus cf. *Crotaphytus reticulatus:* 2
Crotaphytus sp.: 10
Phrynosomatidae
Cophosaurus texanus: 9
Phrynosoma cornutum: 3, 9
Holbrookia sp.: 3, 4
Sceloporus near *Sceloporus olivaceus:* 2
Sceloporus poinsettii: 9
Sceloporus undulatus: 3, 7, 9
Sceloporus near *Sceloporus undulatus:* 2
Sceloporus variabilis: 9
Sceloporus sp.: 10
cf. Iguanidae sp.: 20
Teiidae
Cnemidophorus cf. *Cnemidophorus gularis:* 5
Cnemidophorus sexlineatus: 10

Cnemidophorus sp.: 9, 10, 19
Scincidae
Eumeces fasciatus: 2
Eumeces cf. *Eumeces fasciatus:* 1
Eumeces obsoletus: 2, 8, 9, 10
Eumeces cf. *Eumeces obsoletus:* 7
Eumeces tetragrammus: 9
Anguidae
Ophisaurus attenuatus: 14
Ophisaurus "giant form": 2
Ophisaurus sp.: 5
Colubridae
Arizona elegans: 5, 6, 7, 9, 10, 11
Bogartophis subocularis: 11
Coluber or *Masticophis:* 2, 3, 4, 5, 6, 7, 9, 10, 11, 19, 20, 25
Coluber contrictor: 9, 8, 14, 18
Diadophis punctatus: 9
Diadophis cf. *Diadophis punctatus:* 11
Elaphe guttata: 2, 8, 9
Elaphe cf. *Elaphe guttata:* 7, 11
Elaphe guttata or *obsoleta:* 19
Elaphe obsoleta: 7
Elaphe sp.: 9, 10
Elaphe or *Pituophis:* 8, 10
Gyalopion canum: 8, 11
Heterodon nasicus: 5, 6, 8, 9, 12, 13
Heterodon cf. *Heterodon nasicus:* 7
Heterodon platirhinos: 9, 20
Heterodon cf. *Heterodon platirhinos:* 7
Heterodon sp.: 1, 4, 19
Hypsiglena torquata: 11
Lampropeltis calligaster: 9, 10
Lampropeltis calligaster or *getula:* 1
Lampropeltis getula: 2, 3, 6, 7, 8, 9, 11
Lampropeltis cf. *Lampropeltis getula:* 19
Lampropeltis mexicana: 11
Lampropeltis triangulum: 2, 5, 6, 7, 8, 9, 10, 11, 19
Lampropeltis sp.: 9
Masticophis sp.: 9
Nerodia cyclopion or *rhombifera:* 8
Nerodia erythrogaster: 3, 11, 19
Nerodia cf. *Nerodia erythrogaster:* 1
Nerodia erythrogaster or *fasciata:* 9
Nerodia erythrogaster or *sipedon:* 5, 6
Nerodia fasciata: 2

Nerodia sp.: 8, 14, 20
Opheodrys aestivus: 4, 5, 10, 11
Opheodrys cf. *Opheodrys aestivus:* 7
Opheodrys sp.: 9, 10
Pituophis melanoleucus: 7, 9, 11, 14, 24
Rhinocheilus lecontei: 1, 3, 9, 11
Salvadora sp.: 8
Sonora semiannulata: 8
Sonora sp.: 11
Storeria cf. *Storeria dekayi:* 4
Storeria sp.: 2
Tantilla sp.: 2, 9, 10, 11
Thamnophis marcianus: 8
Thamnophis cf. *Thamnophis marcianus:* 11
Thamnophis marcianus or *radix:* 6
Thamnophis proximus: 6, 8, 9
Thamnophis cf. *Thamnophis proximus:* 3, 4, 5
Thamnophis proximus or *sirtalis:* 19
Thamnophis sirtalis: 4, 9
Thamnophis cf. *Thamnophis sirtalis:* 8
Thamnophis sp.: 1, 2, 4, 7, 9, 11, 12, 13, 14, 15, 19, 25
Tropidoclonion lineatum: 1, 2, 8, 12
Virginia striatula: 5
Virginia cf. *Virginia striatula:* 8

Elapidae
Micrurus fulvius: 2, 9

Viperidae
Agkistrodon contortrix: 6, 8, 9
Agkistrodon piscivorus: 5, 9
Agkistrodon sp.: 7, 9
Crotalus atrox: 2, 3, 8, 10, 24
Crotalus horridus: 2
Crotalus sp.: 1, 5, 6, 7, 9, 10, 11, 19, 20

Alligatoridae
Alligator mississippiensis: 3, 18, 20, 22
Alligator sp.: 1

Rocky Mountain/Great Basin Region

The Rocky Mountain/Great Basin region of this book contains a large variety of vertebrate localities ranging from caves to various low-energy aquatic sites. Some of the herpetofaunas of these sites are very diverse, whereas others have sparse faunal lists compared to other regions. The following states are included in this section (Fig. 29):

Colorado
New Mexico
Utah
Arizona
Nevada

Sites of Special Interest

1. Curtis Ranch Fauna, Arizona

The Curtis Ranch fauna in Cochise County in the southeastern corner of Arizona is the only early Irvingtonian herpetofauna (IrI; Woodburne, 1987) in the Rocky Mountain/Great Basin region. The animals were probably preserved in pools or low-energy streams. An extinct species of kingsnake, *Lampropeltis intermedius* Brattstrom, 1955a, was described from this site on the basis of a trunk vertebra (Fig. 38) that differed from other species in the genus in a combination of characters on the zygosphenal area and on the bottom of the centrum (Brattstrom, 1955a, pp. 152 and 153 and Fig. 1).

Reference. Brattstrom (1955a).

Herpetofauna. *Ambystoma tigrinum*, (?) *Gopherus* sp., *Coluber constrictor*, *Lampropeltis intermedius* **(type locality)**, *Nerodia* sp., *Thamnophis* sp., and *Crotalus lepidus*.

2. Hansen Bluff Fauna, Colorado

The Hansen Bluff fauna of Alamosa County in south-central Colorado is significant because it occurs at a high elevation (2,300 m) and is associated with

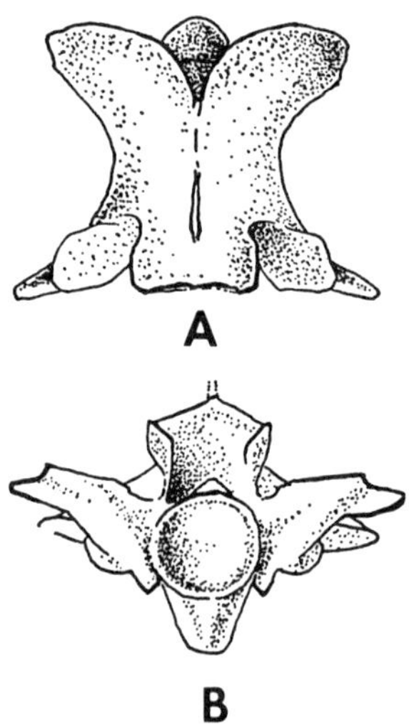

FIGURE 38. Holotype vertebra (California Institute of Technology Number 5121) of *Lampropeltis intermedius* Brattstrom from the Irvingtonian Curtis Ranch fauna of Cochise County, Arizona. (A) Dorsal; (B) anterior. [From Brattstrrom, 1955a]

the Bishop Ash and the Brunhes-Matuyama magnetic boundary. The fauna accumulated in quiet water such as a lake or a pond; in fluvial, low velocity currents; and as fluvial stream-channel deposits. The presence of the voles *Microtus paroperarius* and *Pitymys meadensis* as well as dates on the ash places the site in the late Irvingtonian (IrII; Repenning, 1987). A new species of *Ambystoma* at the site has yet to be formally described.

Reference. Rogers et al. (1985).

Herpetofauna. *Ambystoma tigrinum*, *Ambystoma* "unpublished new species," *Scaphiopus bombifrons*, *Bufo cognatus*, *Bufo woodhousii*, *Bufo* sp., *Pseudacris triseriata*, *Rana blairi* or *pipiens*, *Rana catesbeiana*, *Rana* sp., *Holbrookia maculata*, *Phrynosoma douglasii*, *Sceloporus undulatus*, *Heterodon nasicus*, *Pituophis melanoleucus*, *Thamnophis elegans*, *Thamnophis sirtalis*, *Thamnophis* sp., and *Crotalus viridis*.

3. U-Bar Cave Fauna, New Mexico

This cave fauna occurs in Hidalgo County in extreme southwestern New Mexico. The fauna occurred as a natural cave accumulation. Fourteen radiocarbon dates, six of them taken from red guano, brackets this fauna between ca. 36,000 and 26,000 ybp in the Wisconsinan. All taxa that were identified from this fauna could be found living in Hidalgo County today.

Reference. Harris (1987).

Herpetofauna. *Ambystoma tigrinum*, *Scaphiopus* sp., *Bufo* sp., Testudines sp., *Crotaphytus collaris*, *Phrynosoma cornutum*, *Phrynosoma douglasii*, *Sceloporus* sp. (large and small), cf. *Urosaurus ornatus*, *Cnemidophorus* sp., (?) *Pituophis* sp., (?) *Salvadora* sp., and *Crotalus* sp.

4. Dry Cave Fauna, New Mexico

The Dry Cave fauna of Eddy County in southeastern New Mexico near Carlsbad Cavern has yielded amphibians and reptiles from two portions of the Wisconsinan glacial age. The vertebrate remains in the cave came from natural accumulations as well as the activities of predators. Unit I is bracketed between 34,000 and 25,000 ybp based on radiocarbon dates, and Unit II (localities 4 and 6 of Holman, 1970) represents a time in the late Wisconsinan, 14,000 ybp, based on a radiocarbon date from woodrat dung in the middle of the layers comprising locality 6. The extinct small tortoise *Geochelone wilsoni* was present in Unit I, but all the other identified herpetological taxa are extant. Most of them live in Eddy County today, but some occur at different elevations. *Pseudacris triseriata*, however, is a northern extralimital species, and *Gopherus agassizii* is a western extralimital form from the Mojave Desert.

References. Holman (1970), Harris (1987).

Herpetofauna. Unit I: Anuran sp., *Geochelone wilsoni*, *Gopherus agassizii*, *Crotaphytus collaris*, (?) *Cophosaurus texanus*, *Phrynosoma cornutum*, *Phrynosoma douglasii*, *Phrynosoma modestum*, and *Crotalus* sp. Unit II: *Ambystoma tigrinum*, *Scaphiopus bombifrons*, *Scaphiopus hammondii*, *Scaphiopus* sp., *Bufo punctatus*, *Bufo woodhousii woodhousii*, *Bufo* sp.,

Pseudacris triseriata, *Rana pipiens* complex, *Crotaphytus collaris*, *Phrynosoma cornutum*, *Phrynosoma douglasii*, *Sceloporus undulatus*, *Elaphe guttata*, *Salvadora* sp., *Thamnophis proximus*, *Thamnophis* sp., *Crotalus atrox*, and *Crotalus* sp.

5. Hidden Cave Fauna, Nevada

This fauna from the LaHonton Basin near Reno in western Nevada is from a natural cave fill. It has been the subject of a comprehensive archaeological report with strict stratigraphic control. A few Pleistocene herpetological records were recorded from late Wisconsinan units bracketed by radiocarbon dates: Unit I, 21,000 to 18,000 ybp; Unit II, 18,000 to 15,000 ybp.

Reference. Mead (1985).

Herpetofauna. Unit I: *Scaphiopus intermontanus*, *Scaphiopus* sp., Lizard sp., *Pituophis melanoleucus*, *Lampropeltis* cf. *Lampropeltis zonata*, and Colubridae sp. Unit II: *Pituophis melanoleucus*.

6. Smith Creek Cave Fauna, Nevada

The Smith Creek Cave fauna from the Snake Range in White Pine County next to the Utah border in eastern central Nevada has recently yielded an extensive herpetofauna from packrat middens. Earlier studies (Brattstrom, 1958b, 1976) has provided a herpetofaunal list from sediments of unknown age from this cave. But the longer, stratigraphically controlled list of Mead et al. (1982) will be recorded here, although it does not duplicate all of Brattstrom's records. The herpetofauna reported by Mead et al. is from a reddish brown silt zone that represents the late Wisconsinan based on several radiocarbon dates on dung, unidentified plants, and wood from packrat middens. All the amphibians and reptiles listed below represent extant taxa, but some forms occur outside the Snake Range today. These include *Bufo boreas*, *Bufo* cf. *Bufo woodhousii*, *Sceloporus magister*, *Phrynosoma douglassii*, and *Lampropeltis triangulum*. *Lampropeltis triangulum* occurs farther west in Utah and outside the Great Basin.

References. Brattstrom, 1958b, 1976, Mead, et al. (1982).

Herpetofauna. *Scaphiopus* cf. *Scaphiopus intermontanus*, *Bufo boreas*, *Bufo* cf. *Bufo woodhousii*, *Bufo* sp., *Rana* sp., *Crotaphytus collaris*, *Crotaphytus* sp., *Gambelia wislizenii*, *Phrynosoma douglasii*, *Phrynosoma platyrhinos*, *Phrynosoma* sp., *Sceloporus graciosus*, *Sceloporus magister*, *Sceloporus occidentalis* or *undulatus*, *Sceloporus* sp., *Cnemidophorus* cf. *Cnemidophorus tigris*, *Hypsiglena torquata*, *Lampropeltis pyromelana*, *Lampropeltis triangulum*, *Pituophis melanoleucus*, *Rhinocheilus lecontei*, *Thamnophis* sp., *Crotalus viridis*.

7. Shelter Cave, New Mexico

This herpetofaunal locality lies in Dona Ana County, which borders Mexico in southwestern New Mexico. The amphibian and reptile fossils occur as an accumulation in a dry cave. Some of the gopher tortoise bones in the cave were burned, suggesting that they may have been eaten by aboriginal people. The assemblage, at least in part, represents the late Wisconsinan based on the presence of extinct mammalian taxa, but it is possible that early Holocene bones are also present. Preservation is so excellent that fragments of skin surround some of the skeletons. All the amphibian and reptiles represent living taxa that might be found in or near the area today, with the exception of *Gopherus agassizii,* which is found farther west.

References. Brattstrom (1964), Van Devender, et al. (1976).

Herpetofauna. *Scaphiopus* cf. *Scaphiopus couchii, Rana* cf. *Rana pipiens* complex, *Gopherus agassizii, Crotaphytus collaris, Phrynosoma cornutum, Eumeces obsoletus, Bogartophis subocularis, Coluber constrictor, Lampropeltis getula, Masticophis flagellum, Pituophis melanoleucus,* and *Crotalus atrox.*

8. Howell's Ridge Cave, New Mexico

Howell's Ridge Cave occurs in Grant County in southwestern New Mexico next to the Arizona border. The cave occurs as a vertical chimney that was filled with fossiliferous matrix to within 1 m of the opening. A very large herpetofauna, derived from packrat middens and from the activities of predators, was recorded from the cave. The species list here comes from the 150- to 180-cm unit of the cave that was radiocarbon dated from 11,541 to 9,899 ybp and thus represents the latest Wisconsinan or perhaps the very earliest Holocene. Other higher layers in the cave range well into the Holocene. All the amphibians and reptiles from the cave represent living taxa, as far as can be determined, and most may be found living in the area today.

Reference. Van Devender and Worthington (1977).

Herpetofauna. *Ambystoma tigrinum, Scaphiopus bombifrons* or *hammondii, Scaphiopus couchii, Scaphiopus hammondii, Scaphiopus* sp., *Bufo* sp., *Hyla* cf. *Hyla arenicolor, Crotaphytus collaris, Cophosaurus texanus, Holbrookia maculata, Phrynosoma cornutum, Phrynosoma douglassii, Phrynosoma modestum, Sceloporus* cf. *Sceloporus clarkii, Sceloporus* cf. *Sceloporus jarrovii, Sceloporus* cf. *Sceloporus magister, Sceloporus* cf. *Sceloporus undulatus, Cnemidophorus* cf. *Cnemidophorus tigris, Cnemidophorus* sp., *Leptotyphlops* sp., *Arizona elegans, Diadophis punctatus, Gyalopion canum, Heterodon nasicus, Hypsiglena torquata, Lampropeltis getula, Lampropeltis* cf. *Lampropeltis pyromelana, Masticophis* sp., *Pituophis melanoleucus, Rhinocheilus lecontei, Salvadora* sp., *Sonora semiannulata, Tantilla* sp., *Thamnophis* cf. *Thamnophis cyrtopsis, Trimorphodon biscutatus, Crotalus* large species, and *Crotalus* small species.

9. Deadman Cave, Arizona

This site is found in Pima County, which abuts the Mexican border in south-central Arizona. The cave is presently entered by an approximately 3- by 5-m entrance. Bones accumulated in the cave mainly through the activities of owl and small mammalian predators. It is estimated that the Deadman Cave fauna assemblage most likely dates between 12,000 and 8,000 ybp, grading across the late Wisconsinan–early Holocene boundary (Mead, et al. 1984). All the amphibian and reptile taxa represent extant forms. All still occur in southern Arizona today, but one amphibian and three reptile species have become locally extirpated.

Reference. Mead et al. (1984).

Herpetofauna. *Scaphiopus couchii*, *Scaphiopus* cf. *Scaphiopus hammondii*, *Bufo punctatus*, *Bufo* cf. *Bufo woodhousii*, *Rana* sp., *Callisaurus draconoides*, *Crotaphytus collaris*, *Cophosaurus texanus*, *Holbrookia maculata*, *Phrynosoma douglasii*, *Phrynosoma modestum*, *Phrynosoma solare*, *Sceloporus* cf. *Sceloporus clarkii*, *Sceloporus* cf. *Sceloporus magister*, *Sceloporus* cf. *Sceloporus undulatus*, *Urosaurus ornatus*, *Cnemidophorus* sp., *Heloderma suspectum*, *Arizona elegans*, *Gyalopion canum*, *Hypsiglena torquata*, *Lampropeltis getula*, *Lampropeltis pyromelana*, *Masticophis* sp., *Pituophis melanoleucus*, *Rhinocheilus lecontei*, *Salvadora* sp., *Trimorphodon biscutatus*, *Crotalus atrox*, and *Crotalus scutulatus*.

10. Gypsum Cave, Nevada

This cave lies in Clark County in the southeastern tip of Nevada near Las Vegas. The site represents a Pleistocene cave den. A date of 11,690 $\pm$ 250 ybp was obtained on sloth dung (Lundelius et al., 1983). The herpetofauna represents taxa that are all living today in southern Nevada.

References. Bratttstrom (1954a, 1958b), Tihen (1962).

Herpetofauna. *Bufo punctatus*, *Rana pipiens* complex, *Gopherus agassizii*, *Sauromalus obesus*, *Crotaphytus collaris*, *Phrynosoma platyrhinos*, *Cnemidophorus tigris*, *Heloderma suspectum*, *Lampropeltis getula*, *Lampropeltis pyromelana*, *Masticophis flagellum*, *Masticophis taeniatus*, *Pituophis melanoleucus*, *Crotalus atrox*, *Crotalus mitchellii*, and *Crotalus viridis*.

Other Important Sites

11. Brown Sand Wedge Site, New Mexico

This site accumulated in a lacustrine setting and dates between 12,000 and 11,000 ybp on the basis of human artifacts (Lundelius et al., 1983). This is an important locality, for it contains the small extinct tortoise *Geochelone wilsoni*.

References. Milstead (1967), Preston (1979).

12. Burnet Cave, Eddy County, New Mexico

The vertebrates of this site originated from a natural cave fill and perhaps from aboriginal activities. A date of about 11,500 ybp is suggested by the presence of a Folsom-type point. Four taxa of lizards were identified from this deposit.

Reference. Rickart (1977).

13. Dark Canyon Cave, Eddy County, New Mexico

The vertebrates from this site originated from a natural cave fill. It is believed to represent the late Pleistocene (Rickart, 1977). The same four taxa of lizards that were identified from Burnet Cave were identified from Dark Canyon Cave (see summary list, this section).

Reference. Rickart (1977).

14. Conkling Cave, Dona Ana County, New Mexico

This cave may have been inhabited by humans, as some of the vertebrate bones were burned. The deposit is believed to represent the late Wisconsinan based on extinct mammalian remains, but it is possible that the early Holocene is represented as well.

References. Brattstrom (1961, 1964), Van Devender et al. (1976).

15. Robledo Cave, Dona Ana County, New Mexico

This cave contains gopher tortoise bones that are charred as well as potsherds and charcoal. The tortoise bones are well-preserved, so that it has been suggested by Van Devender et al. (1976) that the site could be Pleistocene in age.

Reference. Van Devender et al. (1976).

16. Wylde Cave, Dona Ana County, New Mexico

This cave is a small bat cave that is assumed to be late Wisconsinan to early Holocene in age.

Reference. Brattstrom (1964).

17. Crystal Ball Cave, Millard County, Utah

The bones in this cave are attributed to packrats and to storm runoff. The age ranges from at least 23,000 radiocarbon years before the present into the Holocene.

Reference. Mead et al. (1989).

18. Rampart Cave, Mohave County, Arizona

Amphibian and reptile bones from Rampart Cave are derived from packrat middens. Radiocarbon dates on twigs, seeds, and mammalian dung on the layers bearing herpetological fossils bracket a time between 19,000 and 9,500 ybp, thus representing late Wisconsinan to very early Holocene times.

Reference. Van Devender et al. (1977).

19. Vulture Canyon, Mohave County, Arizona

The herpetologcal remains from this site came from packrat middens in a cave. The amphibian and reptile fauna is bracketed between radiocarbon dates of 13,000 and 12,000 ybp, obtained from seeds and twigs and thus representing the late Wisconsinan.

Reference. Van Devender et al. (1977).

20. Emery Falls Canyon, Mohave County, Arizona

Amphibian and reptile remains from this site come from packrat middens and carnivore activities in a cave. Pleistocene localities in the cave are bracketed between 11,000 and 10,000 ybp on the basis of radiocarbon dates on leaves and twigs, and thus the site represents the very late Wisconsinan.

Reference. Van Devender et al. (1977).

21. Wolcott Peak, Pima County, Arizona

Herpetological remains from this packrat midden site come from two localities bracketed in time between about 14,550 and 5,000 ybp; thus a late Wisconsinan to mid-Holocene designation would apply.

Reference. Van Devender and Mead (1978).

22. Brass Cap Point, Yuma County, Arizona

The amphibian and reptile remains from this site come from a single packrat midden locality with a late Wisconsinan radiocarbon date of 11,450 $\pm$ 400 ybp.

Reference. Van Devender and Mead (1978).

23. New Water Mountains, Yuma County, Arizona

The New Water Mountains packrat midden herpetological localities are bracketed between radiocarbon dates of ca. 12,000 and 8,000 ybp; thus a late Wisconsinan to early Holocene designation would apply.

Reference. Van Devender and Mead (1978).

24. Welton Hills, Yuma County, Arizona

The Welton Hills packrat midden localities that enter the Pleistocene are bracketed from ca. 11,000 to ca. 9,000 ybp based on radiocarbon dates. Therefore, the herpetological remains listed from this site would be designated as very late Plesitocene to very early Holocene in age.

Reference. Van Devender and Mead (1978).

25. Glendale Fauna, Clark County, Nevada

The Glendale fauna occurred in a site that represents a pluvial lake or a marsh along a permanent river. The site is thought to represent the "late Rancholabrean" on the basis of a radiocarbon date on charcoal 40 cm below the turtles and associated mammals (Van Devender and Tessman, 1975). The site is important because *Chelydra serpentina* occurs there, at least 650 miles west of its westernmost occurrence today.

Reference. Van Devender and Tessman (1975).

26. Snake Creek Burial Cave, White Pine County, Nevada

This cave formed a natural trap for vertebrates. Reptiles and amphibians range from late Rancholabrean to early Holocene in age at this locality.

Reference. Mead et al. (1989).

Summary of Herpetological Taxa*

Ambystomatidae
Ambystoma tigrinum: 1, 2, 3, 4-II, 8
Ambystoma "unpublished new species": 2
Pelobatidae
Scaphiopus bombifrons: 2, 4-II
Scaphiopus bombifrons or *hammondii:* 8
Scaphiopus couchii: 8, 9
Scaphiopus cf. *Scaphiopus couchii:* 7
Scaphiopus hammondii: 4-II, 8
Scaphiopus cf. *Scaphiopus hammondii:* 9
Scaphiopus intermontanus: 5-I
Scaphiopus cf. *Scaphiopus intermontanus:* 6
Scaphiopus sp.: 3, 4-II, 5-I, 8
Bufonidae
Bufo boreas: 6
Bufo cognatus: 2
Bufo punctatus: 4-II, 9, 10, 21
Bufo woodhousii woodhousii: 4-II
Bufo woodhousii: 2
Bufo cf. *Bufo woodhousii:* 6, 9
Bufo sp.: 2, 3, 4-II, 6, 8
Hylidae
Hyla cf. *Hyla arenicolor:* 8
Pseudacris triseriata: 2, 4-II
Ranidae
Rana blairi or *pipiens:* 2
Rana catesbeiana: 2
Rana pipiens complex: 4-II, 10
Rana cf. *Rana pipiens* complex: 7
Rana sp.: 2, 9, 6
Anuran sp.: 4-I
Chelydridae
Chelydra serpentina: 25
Emydidae
Terrapene carolina ssp.: 11
Terrapene ornata: 11
Testudinidae
Geochelone wilsoni: 4-I, 11
Gopherus agassizii: 4-I, 7, 10, 14, 15, 18, 19, 24
(?) *Gopherus* sp.: 1

---

*Numerals in summary correspond to sites described in this section.

Testudines sp.: 3
Iguanidae
Callisaurus draconoides: 9
Sauromalus obesus: 10, 18, 19, 20, 21
Crotaphytidae
Crotaphytus collaris: 3, 4-I, 4-II, 6, 7, 8, 9, 10, 12, 13, 16, 20, 21
Crotaphytus cf. *Crotaphytus collaris:* 17
Crotaphytus sp.: 6, 17
Gambelia wislezinii: 6, 17
Phrynosomatidae
Cophosaurus texanus: 8, 9
(?) *Cophosaurus texanus:* 4-I
Holbrookia maculata: 2, 8, 9
Phrynosoma cornutum: 3, 4-I, 4-II, 7, 8, 12
Phrynosoma douglassii: 2, 3, 4-I, 4-II, 6, 8, 9, 12, 13, 17, 26
Phrynosoma modestum: 4-I, 8, 9
Phrynosoma platyrhinos: 6, 10, 17
Phrynosoma solare: 9
Phrynosoma sp.: 6
Sceloporus cf. *Sceloporus clarkii:* 8, 9
Sceloporus graciosus: 6, 17
Sceloporus cf. *Sceloporus jarrovii:* 8
Sceloporus magister: 6
Sceloporus cf. *Sceloporus magister:* 8, 9, 18, 20, 22, 23, 24
Sceloporus occidentalis or *undulatus:* 6, 17, 26
Sceloporus undulatus; 2, 4-II
Sceloporus cf. *Sceloporus undulatus:* 8, 9, 18, 19, 20
Sceloporus sp.: 3, 6, 12, 13, 17, 20, 21
Urosaurus cf. *Urosaurus graciosus:* 24
Urosaurus ornatus: 9
cf. *Urosaurus ornatus:* 3
Uta stansburiana: 17, 20, 22, 24
Lizard sp.: 5-I
Gekkonidae
Coleonyx variegatus: 24
Teiidae
Cnemidophorus tigris: 10
Cnemidophorus cf. *Cnemidophorus tigris:* 6, 8, 17, 19, 21
Cnemidophorus sp.: 3, 8, 9, 12, 13, 20, 21
Scincidae
Eumeces obsoletus: 7, 16
Helodermatidae
Heloderma suspectum: 9, 10
Leptotyphlopidae
Leptotyphlops cf. *Leptotyphlops humilis:* 21

Leptotyphlops sp.: 8
Colubridae
Arizona elegans: 8, 9, 23, 24
Bogartophis subocularis: 7, 14, 16
Chionactis occipitalis: 21, 24
Coluber constrictor: 1, 7, 14
Diadophis punctatus: 8
Elaphe guttata: 4-II
Gyalopion canum: 8, 9
Heterodon nasicus: 2, 8
Hypsiglena torquata: 6, 8, 9, 19, 21, 23
Lampropeltis getula: 7, 8, 9, 10, 18
Lampropeltis intermedius: 1
Lampropeltis pyromelana: 6, 9, 10
Lampropeltis cf. *Lampropeltis pyromelana:* 8
Lampropeltis triangulum: 6
Lampropeltis cf. *Lampropeltis zonata:* 5-I
Lampropeltis sp.: 17, 26
Masticophis flagellum: 7, 10
Masticophis taeniatus: 10
Masticophis sp.: 8, 9, 21
Nerodia sp.: 1
Phyllorhynchus decurtatus: 23, 24
Phyllorhynchus cf. *Phyllorhynchus decurtatus:* 21
Phyllorhynchus sp.: 21
Pituophis melanoleucus: 2, 5-I, 5-II, 6, 7, 8, 9, 10, 14, 18, 20, 21, 24, 26
(?) *Pituophis* sp.: 3
Rhinocheilus lecontei: 6, 8, 9, 18, 21, 23
Salvadora sp.: 4-II, 8, 9
(?) *Salvadora* sp.: 3
Sonora semiannulata: 8, 20, 23
Sonora sp.: 11
Tantilla sp.: 8
Thamnophis cf. *Thamnophis cyrtopsis:* 8
Thamnophis elegans: 2
Thamnophis proximus: 4-II
Thamnophis sirtalis: 2
Thamnophis sp.: 1, 2, 4-II, 6
cf. *Thamnophis* sp.: 17
Trimorphodon biscutatus: 8, 9, 22, 23, 24
Colubridae sp.: 5-I
Viperidae
Crotalus atrox: 4-II, 7, 9, 10, 14
Crotalus cerastes: 24

Crotalus lepidus: 1
Crotalus mitchellii: 10
Crotalus mitchellii or *viridis:* 20
Crotalus scutulatus: 9
Crotalus viridis: 2, 6, 10, 26
Crotalus large species: 8
Crotalus small species: 8
Crotalus sp.: 3, 4-I, 4-II, 19

Pacific Coast Region

The Pacific Coast region contains several kinds of Pleistocene sites including caves, tar pits, and fluvial and lacustrine deposits, but the actual number of herpetological localities and the diversity of herpetological species is more limited than one might imagine. The following state is included in this section (Fig. 29):

California

Sites of Special Interest

1. Costeau Pit, California

The Costeau Pit of Orange County south of Los Angeles has yielded an interesting herpetofauna that was originally described by Miller (1971) and modified by Hudson and Brattstrom (1977). The bones were evidently deposited in a low-energy aquatic situation. The site is said to be Rancholabrean with an estimated age of ±40,000 ybp (Hudson and Brattstrom, 1977). All the taxa represent extant forms and all occurred in the vicinity of the deposit in historic times except for *Gopherus,* which occurs east of the site today in desert areas.

References. Miller (1971), Hudson and Brattstrom (1977).

Herpetofauna. *Aneides lugubris*, *Bufo boreas*, *Rana aurora*, *Clemmys marmorata*, *Gopherus* sp., *Sceloporus* sp., *Cnemidophorus* sp., *Pituophis melanoleucus*, and *Crotalus viridis*.

2. Newport Beach Mesa Site, California

This site also occurs in Orange County, California, south of Los Angeles. The bones were taken from a clay matrix that was the result of stream deposition. The site occurred sometime in late Pleistocene Rancholabrean times (Hudson and Brattstrom, 1977). All the taxa are extant and all occur in the area today. The site is noteworthy as the first reported locality for the salamamder *Aneides lugubris*.

Reference. Hudson and Brattstrom (1977).

Herpetofauna. *Aneides lugubris*, *Bufo boreas*, *Pseudacris regilla*, *Rana aurora*, *Lampropeltis getula*, *Masticophis flagellum*, *Thamnophis* sp., and *Crotalus viridis*.

3. Rancho La Brea, California

The Rancho La Brea site of Los Angeles County, within the city limits of Los Angeles, is one of the most famous Pleistocene sites, if not the most famous Pleistocene site in the world. It is also by far the largest and most diverse herpetofaunal assemblage in the entire Pacific Coast region. The site is a fluvial one in which the vertebrate fauna was trapped in asphalt-impregnated streams and ponds, giving us an invaluable window of life in the late Pleistocene.

The herpetofauna of the Rancho La Brea site comes from localities that range from 9,000 to 27,000 radiocarbon years from the late medial Wisconsinan into the earliest Holocene. The amphibians and reptiles of this site are now all considered to represent living taxa, most of which occur in the area today with the exception of the possible occurrence of an extinct tortoise "? *Geochelone* sp." (Fay, 1991) and *Sceloporus magister,* which occurs in more xerophytic lands to the east at present. This is in sharp contrast to the mammalian fauna, which has many extinct species.

References. Brattstrom (1953b, 1954b, 1958a), Hecht (1956), Tihen (1962), Fay (1991), LaDuke (1991a, 1991b), Bell and Dundas (1993).

Herpetofauna. *Aneides lugubris*, *Bufo boreas haplophilus*, *Pseudacris* sp., *Rana aurora*, *Clemmys marmorata*, (?) *Geochelone* sp., *Phrynosoma coronatum*, *Sceloporus magister*, *Sceloporus occidentalis*, *Uta stansburiana*, *Cnemidophorus tigris*, *Eumeces skiltonianus*, *Gerrhonotus multicarinatus*, *Arizona elegans*, *Coluber constrictor*, *Coluber* or *Masticophis*, *Diadophis punctatus*, *Hypsiglena torquata*, *Lampropeltis getula*, *Masticophis lateralis*, *Pituophis melanoleucus*, *Rhinocheilus lecontei*, *Tantilla* sp., *Thamnophis couchii* complex, *Thamnophis sirtalis*, *Thamnophis* sp., and *Crotalus viridis*.

4. McKittrick Asphalt, California

This site lies in Kern County in southern California north of Los Angeles. The fauna of this site was thought to be trapped in asphalt at a water hole. Based on the faunal assemblage this site is thought to represent late Wisconsinan times. There are no surprises in this late Pleistocene herpetofauna, as all the identified taxa represent extant species and can be found in the area today except for the tortoise *Gopherus agassizii*, which occurs farther to the west in the Mojave Desert.

References. Miller (1942), Brattstrom (1953a, 1954b).

Herpetofauna. *Clemmys marmorata*, *Gopherus agassizii*, *Gambelia wislizenii*, *Sceloporus occidentalis*, *Cnemidophorus tigris*, *Eumeces gilberti*, *Lampropeltis getula*, *Pituophis melanoleucus*, and *Crotalus viridis*.

5. Potter Creek Cave, California

The Potter Creek Cave site lies in Shasta County in central northern California. The site represents a cave den. A Wisconsinan age fauna is suggested by the

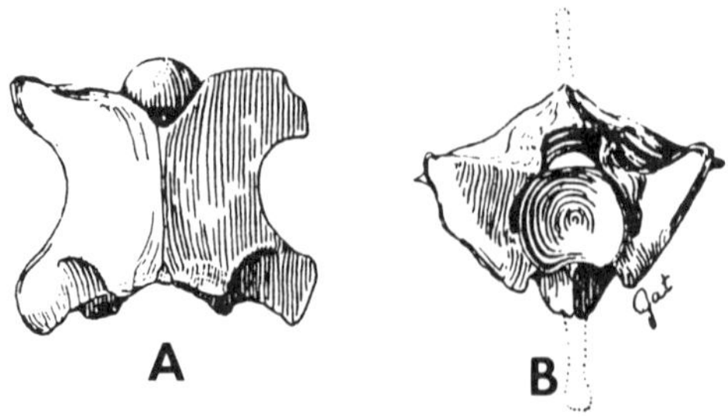

FIGURE 39. Holotype vertebra (University of California Museum of Paleontology Number 3129) of *Crotalus potterensis* Brattstrom from the Rancholabrean Potter Creek site, Shasta County, California. (A) Dorsal; (B) anterior. [From Brattstrom, 1953a]

faunal assemblage (Lundelius et al., 1983). The site is noteworthy because it has produced the extinct Pleistocene rattlesnake *Crotalus potterensis* Brattstrom (1953a). This questionable species was described on the basis of a vertebra (Fig. 39) and will be further discussed in Chapter 8 of this book.

References. Brattstrom (1953a, 1954a, 1958a), Tihen (1962).

Herpetofauna. *Bufo boreas*, *Clemmys* sp., *Crotalus potterensis* **(type locality)**.

Other Important Sites

6. Hawver Cave, El Dorado County, California

The bones from this site were mainly derived from predators in the cave. It is believed to represent the Wisconsinan based on the faunal assemblage (Lundelius et al., 1983).

References. Brattstrom (1953a, 1954b).

7. Kings Canyon Biota, South-Central Sierra Nevada, California

This biota was recovered from a packrat midden. The fauna is thought to represent the late Pleistocene based on infinite radiocarbon dates.

Reference. Mead et al. (1985).

8. Mescal Cave, San Bernardino County, California

Vertebrate fossils from this site represent the natural accumulation in a cave. The site is thought to represent the late Pleistocene, but there may be some Holocene species as well.

Reference. Brattstrom (1958a).

9. Schuiling Cave, San Bernardino County, California

This cave fauna is the result of predator activity in a den as well as flooding. It is thought to represent the late Pleistocene on the basis of its faunal assemblage (Lundelius et al., 1983).

Reference. Brattstrom (1958a).

10. Carpenteria, Santa Barbara County, California

The fauna at the Carpenteria site was trapped in an asphalt deposit. The site is thought to be late Wisconsinan in age based on the faunal assemblage (Lundelius et al., 1983).

Reference. Brattstrom (1955b).

11. Falling Arches Site, San Bernardino County, California

This fauna originated from packrat middens and a record of *Sauromalus obesus* comes from a single late Wisconsinan locality with a radiocarbon date of 11,650 ± 190 ybp.

Reference. Van Devender and Mead (1978).

12. Redtail Peaks Site, San Bernardino, California

These vertebrate remains originated from packrat middens at four localities bracketed between 14,000 and 9,000 ybp and thus represent late Wisconsinan to very early Holocene times.

Reference. Van Devender and Mead (1978).

13. Tunnel Ridge Site, San Bernardino County, California

The herpetofauna at this site originated from packrat middens and came from a single very late Wisconsinan locality radiocarbon dated 10,330 ± 300 ybp.

Reference. Van Devender and Mead (1978).

14. Whipple Mountain Site, San Bernardino County, California

This fossil assemblage was recovered from packrat middens from two very late Wisconsinan localities that had radiocarbon dates that bracketed the fauna between about 10,500 and 10,000 ybp.

Reference. Van Devender and Mead (1978).

Summary of Herpetological Taxa*

Plethodontidae
- *Aneides lugubris:* 1, 2, 3
- *Hydromantes* sp.: 7

Bufonidae
- *Bufo boreas haplophilus:* 3
- *Bufo boreas:* 1, 2, 5
- *Bufo punctatus:* 12, 13

Hylidae
- *Pseudacris regilla:* 2
- *Pseudacris* sp.: 3

Ranidae

---

*Numerals in summary correspond to sites described in this section.

Rana aurora: 1, 2, 3
Emydidae
Clemmys marmorata: 1, 3, 4, 10
Clemmys sp.: 5
Testudinidae
(?) *Geochelone* sp.: 3
Gopherus agassizii: 4, 8, 9, 14
Gopherus sp.: 1
Iguandiae
Sauromalus obesus: 9, 11
Crotaphytidae
Crotaphytus sp.: 8
Gambelia wislizenii: 4, 13
Phrynosomatidae
Phrynosoma coronatum: 3
Sceloporus graciosus: 6
Sceloporus magister: 3
Sceloporus occidentalis: 3, 4, 8, 10
Sceloporus cf. *Sceloporus occidentalis:* 7
Sceloporus sp.: 1
Uta stansburiana: 3
Teiidae
Cnemidophorus tigris: 3, 4
Cnemidophorus cf. *Cnemidophorus tigris:* 8, 13
Cnemidophorus sp.: 1
Scincidae
Eumeces gilberti: 4
Eumeces skiltonianus: 3
Anguidae
Gerrhonotus multicarinatus: 3, 6, 10
Gerrhonotus cf. *Gerrhonotus multicarinatus:* 7
Boidae
Lichanura trivirgata: 12
Colubridae
Arizona elegans: 3
Chionactis occipitalis: 12, 14
Coluber constrictor: 3, 10
Coluber or *Masticophis:* 3
Diadophis punctatus: 3
Hypsiglena torquata: 3, 12, 13, 14
Lampropeltis getula: 2, 3, 4, 12, 13
Masticophis flagellum: 2
Masticophis lateralis: 3
Pituophis melanoleucus: 1, 3, 4
Rhinocheilus lecontei: 3, 13

Tantilla sp.: 3
Thamnophis couchii: complex 3
Thamnophis sirtalis: 3
Thamnophis sp.: 2, 3, 7, 10

Viperidae

Crotalus cf. *Crotalus atrox:* 13
Crotalus potterensis: 5
Crotalus viridis: 1, 2, 3, 4, 6, 8
Crotalus sp.: 12

7

HERPETOLOGICAL POPULATION ADJUSTMENTS DURING THE NORTH AMERICAN PLEISTOCENE

Two overriding events took place in vertebrate populations during the Pleistocene: (1) range adjustments due to the obliteration of habitats by glacial advances and altered climatic patterns, and (2) widespread extinctions. This chapter will deal with both of these population adjustments in a general way, and the following chapter will specifically address extinction. The complete faunal list of the localities discussed in the present chapter appears in the preceding chapter.

Range adjustments in herpetological populations may be divided into three major movements: (1) southward movement of northern populations during the glacial advances and the reinvasion of their former ranges by these populations during glacial retreats, (2) the altitudinal range adjustments made in mountainous regions, and (3) range adjustments made by certain species in areas south of the glaciated areas.

It is very important to note the third topic at this time, for not many years ago it would have not been an important consideration. The classic idea was that advancing and retreating glaciers altered climatic isotherms in such a uniform way that communities were uniformly displaced southward during glacial advances and uniformly advanced northward during glacial retreats. In other words, uniform bands of tundra, taiga, and deciduous forest with their component animal communities were envisioned as having moved downward during glacial advances and upward during glacial retreats. This classic **stripe hypothesis** was a great oversimplification.

Invoking the stripe hypothesis, many herpetologists believed that climates in the southern United States during glacial advances were much colder during the Pleistocene than they actually were. For instance, some scientists in the 1950s believed that the Pleistocene climate in northern Florida during the glacial periods was similar to modern Virginia. Moreover, herpetologists believed that Mexico became a refugium for many species in the western United States.

Today, based on the study of scores of Pleistocene faunas, we know that very few—if any—"southern" herpetological species moved farther south during the Pleistocene. Other oversimplified explanations of zoogeographic and speciation events have also been the outgrowth of the stripe hypothesis.

The late 1950s marked the beginning of the end of the stripe hypothesis, for it quickly became apparent that many Pleistocene vertebrate species were extralimital northern and southern forms coexisting with a fauna that would be analogous to the one that would occur in that area today. These communities finally became known as **disharmonious communities** (e.g., Lundelius et al., 1983).

To account for disharmonious communities the idea of colder communities far south of the glacial boundaries had to be abandoned. The only parsimonious explanation for the occurrence of northern and southern extralimital species exisitng together in the same faunas was that cooler summers allowed for the presence of the northern species and warmer winters allowed for the presence of the southern ones. This idea has become known as the **Pleistocene climatic equability hypothesis**.

Other studies (e.g., Graham, 1976, 1979; Graham and Lundelius, 1984) grew out of the contemplation of disharmonious communities and the Pleistocene equability hypothesis; it was pointed out that each species of plant or animal adjusts in its own individual way to glacial and interglacial events and that it was naive to assume that entire biomes would march in place from north to south (or east to west) and back again.

Conversely, it appears that certain northern species forced southward by the advancing glaciers were able to adjust to certain situations in the newly invaded southern land and that others would not be able to adjust. We have seen in this book (Chapter 6) that herpetological species made range adjustments not only from north to south but from east to west, especially in the midcontinental region. All this then would lead to a **mosaic** community composed of the original fauna plus the surviving extralimital invaders. This reasoning has lead to the so-called **plaid hypothesis**.

In summary, it is now known that many Pleistocene vertebrate communities south of the glaciated areas were mosaic ones with extralimital invading species superimposed on the original fauna, and it is believed that these communities existed in a climate that was much more equable than it is in these areas today. These concepts will be viewed in the light of Pleistocene amphibian and reptile populations in the following pages of this chapter.

Northeastern/Great Lakes Region: Reoccupation of Glaciated Areas

Range adjustments in this region were the most dramatic of any herpetofaunal region covered in this book because of the penetration of the Laurentide Pleistocene ice sheet deep into the heart of the United States (Fig. 1). The Pleistocene must have been a time of relatively constant range adjustments for herpetological species in the region (Fig. 2), but unfortunately we do not have any sizable herpetological faunas until late in the Wisconsinan and these are very few.

A key fossil site in earlier Pleistocene times is the Hopwood Farm locality in Montgomery County in south-central Illinois. This site yielded the remains of an extinct giant land tortoise, *Geochelone crassiscutata,* which occurred in the Sangamonian horizon of the deposit. The pollen data of the Hopwood Farm site suggests that the vegetation at the time the giant tortoise lived was mainly of grasses and herbaceous plants. The presence of the giant tortoise strongly suggests that temperatures above freezing might have occurred throughout the year during the part of the Sangamonian represented by the tortoise remains (Hibbard, 1960; King and Saunders, 1986). By additional implication, one would expect a very diverse herpetofauna for the area during this part of Sangamonian time.

In the late Wisconsinan the Laurentide ice sheet penetrated deep into the central Great Lakes region to almost identical limits in central Illinois, Indiana, and Ohio about 21,000 to 20,000 ybp (Fig. 1). This eliminated all the herpetological habitat in Michigan, as well as in northeastern Illinois and northern Indiana and Ohio. Other late advances occurred, but none of these reached as far as the 21,000 to 20,000 ybp advance (Fig. 2). The final retreat of the Laurentide ice sheet began about 15,000 ybp.

An interesting glimpse of the herpetological life in southwestern Indiana 15,000 to 14,000 ybp (Holman, 1992a; Holman and Richards, 1993) is provided by the Prairie Creek D herpetofauna of Daviess County (Fig. 40). At Prairie Creek D time the ice margin may have been as close as the northeastern tip of Indiana or as distant as eastern-central lower Michigan (Fig. 2). Most of the reptiles and some of the amphibians of this herpetofauna must have invaded southwestern Indiana rather rapidly from more southern areas, because during the 21,000 to 20,000 ybp interval the ice margin was as near as 95 km from the Prairie Creek D site, south of Indianapolis in northern Morgan and Johnson counties.

One would expect the vegetation at the time of the 21,000 to 20,000 ybp advance, and even as late as the 14,800 ybp advance (Fig. 2), to have been boreal forest (Delcourt, Delcourt, and Webb, 1983), an environment that would be habitable to only a relatively few cold-adapted species. The Prairie Creek D herpetofauna consists of one salamander, five anurans, seven turtles, and two snakes.

Several of these species would have perished in a boreal forest habitat and thus they probably did not reinvade southwestern Indiana until after the withdrawal of the 14,800 ybp ice. These species are *Sternotherus odoratus, Graptemys* sp., *Pseudemys* sp., *Apalone spinifera,* and *Nerodia erythrogaster.* These turtles and snakes are of considerable importance from a paleoecological standpoint, since all of them are not only typical of southwestern Indiana today but are species of common and typical occurrence in the southeastern and south-central United States today (Conant and Collins, 1991).

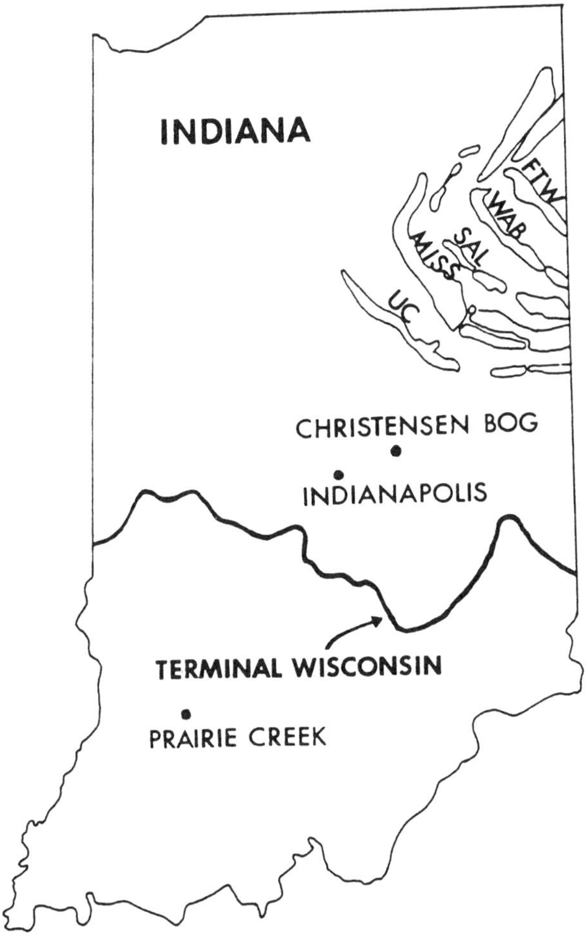

FIGURE 40. Location of the Prairie Creek D and Christensen Bog sites in relation to Wisconsinan moraines in Indiana. Moraines: FTW = Fort Wayne; WAB = Wabash; SAL = Salamonie; MISS = Mississinewa; and UC = Union City. [From Holman, 1992a]

A limited but important look at herpetological life in central Indiana 13,000 to 12,000 ybp is provided by the Christensen Bog mastodon site bone bed Qmp3 zone, Hancock County, Indiana (Graham et al., 1983). This site lies just southwest of the Union City moraine, which is correlated with the 14,000 ybp ice advance (Fig. 40). Thus, one would expect Hancock County, Indiana, would have been tundralike at this date and practically uninhabitable for any herpetological species. But by Christensen Bog time, when the ice front was probably situated in the northern part of lower Michigan and in southern Ontario (Fig. 2), a frog (*Rana pipiens* complex) and three species of turtles (*Chelydra serpentina, Chrysemys picta, Apalone spinifera*) inhabited the area.

Turtles are extremely important members of Pleistocene vertebrate faunas in northern latitudes. The critical time in the life cycle of northern hemisphere turtles is the number of warm days in the summer for egg incubation. A mean July temperature well in excess of 18°C appears to be necessary for the European pond tortoise *Emys orbicularis* to breed successfully in Europe (Stuart, 1979). By inference, *Chrysemys picta* (a cold-adapted North American species that reaches the farthest north of any turtle in the North American continent today) specimens imported to seemingly ideal habitats in southern England have not been able to establish viable populations because the temperatures there are not warm enough for the successful hatching of eggs. Therefore, the fact that turtles occur in the Christensen Bog fauna indicates the probability that summers would have had as many warm days as in the areas where these species occur today.

The Sheriden Pit site of Wyandot County in northwestern Ohio (Holman, 1994) is very important in that it has yielded by far the largest Pleistocene herpetofauna in the Great Lakes Basin proper. A radiocarbon date of ca. 11,000 ybp has been obtained on charcoal from the top layer of this site, thus the lower ages could be older. Nevertheless, fifteen herpetological species have been already identified from this site, all of which could have found near the cave within historic times. This indicates that the herpetofauna must have been mainly reestablished in northwestern Ohio by late Wisconsinan times.

A limited but important indication of herpetological life in the Great Lakes region is provided by the Clark Kettle Bog site in Darke County in west-central Ohio (Holman, 1986b). Two species of turtles, *Chelydra serpentina* and *Chrysemys picta,* both cold-adapted species, existed about 11,000 to 10,000 ybp at this site.

Holman (1992b) developed a model for herpetological reoccupation of postglacial Michigan based on an interpretation of geological data, paleobotanical and paleovertebrate assemblages, and ecological tolerances of the modern herpetofauna. Three categories of invading species were proposed: primary invaders, whose ecological tolerances included coniferous forest and in one case, *Rana sylvatica,* tundra areas; (2) secondary invaders adapted for mixed coniferous–broadleaf situations; and (3) tertiary invaders adapted to exist in broadleaf forest areas (Tables 1–3).

As one might expect, most of the tertiary invaders occur today in isolated or peripheral populations, and some represent endangered species or those of special concern. Two major routes of entry were suggested (Fig. 41): (1) a wide corridor through Indiana and Ohio into the southern peninsula of Michigan; and (2) a somewhat narrower corridor through Wisconsin into the Upper Peninsula. Lesser invasions could have come from Ontario into either one or both peninsulas, but it was suggested that few, if any, species were able to invade the Upper Peninsula from the Lower Peninsula across the Straits of Mackinac.

TABLE 1. Herpetological Taxa Considered to be Primary Invaders of Post-glacial Michigan

Amphibians	Reptiles
Ambystoma laterale complex	*Chelydra serpentina*
Ambystoma maculatum	*Chrysemys picta belli*
Ambystoma tigrinum	*Chrysemys picta marginata*
Hemidactylium scutatum	*Clemmys insculpta*
Necturus maculosus	*Emydoidea blandingii*
Notophthalmus viridescens	*Eumeces fasciatus*
Plethodon cinereus	*Diadophis punctatus*
Bufa americanus	*Elaphe vulpina vulpina*
Hyla versicolor	*Nerodia sipedon*
Pseudacris crucifer	*Opheodrys vernalis*
Pseudacris triseriata	*Storeria occipitomaculata*
Rana catesbeiana	*Thamnophis sirtalis*
Rana clamitans	
Rana palustris	
Rana pipiens	
Rana septentrionalis	
Rana sylvatica	

Note: These species may be found in coniferous forest areas today or, in one case (*Rana sylvatica*), in tundra.

TABLE 2. Herpetological Taxa Considered to be Secondary Invaders of Post-glacial Michigan

Amphibians	Reptiles
Acris crepitans	Clemmys guttata
Bufo woodhousii	Graptemys geographica
Hyla chrysoscelis	Terrapene carolina
	Apalone spinifera
	Coluber constrictor
	Elaphe vulpina gloydi
	Heterodon platirhinos
	Lampropeltis triangulum
	Regina septemvittata
	Sistrurus catenatus
	Storeria dekayi
	Thamnophis butleri
	Thamnophis sauritus

Note: These species extend into the mixed coniferous–broadleaf forest areas of the Lower Peninsula of Michigan today but do not occur in the Upper Peninsula except for a very isolated population of *Coluber constrictor* in the extreme southwestern part of the Upper Peninsula.

TABLE 3. Herpetological Taxa Considered to be Tertiary Invaders of Postglacial Michigan

Taxon	Comment
Amphibians	
Ambystoma opacum	SW peripheral
Ambystoma texanum	SE peripheral
Siren intermedia	SW isolated population
Reptiles	
Sternotherus odoratus	Ranges north to Mason-Quimby Line
Trachemys scripta	Widespread introduced populations
Cnemidophorus sexlineatus	One population in Tuscola County
Clonophis kirtlandii	Endangered, peripheral
Elaphe obsoleta	Special concern, SW
Nerodia erythrogaster	Endangered, peripheral

Note: These species are all confined to the broadleaf area south of the Mason-Quimby Line.

It was suggested that *Rana sylvatica,* which is able to tolerate tundra conditions today, may have been the first primary species to invade postglacial Michigan as the great mass of sterile, glacially derived mud and sand gave way to tundra vegetation between 14,800 and 13,000 years ago. Other primary (coniferous tolerant) forms probably invaded southern Michigan about 12,500 ybp as first spruce forests developed and later (ca 11,800 ybp) jack and red pine communities began to develop. On the other hand, primary species probably did not reach the Upper Peninsula until at least 9,900 ybp, as even during that time glacial ice rimmed the northern border of the area.

Secondary herpetological invaders are thought to have probably begun to enter southern Michigan about 10,600 ybp as mixed coniferous–broadleaf forests began to develop in the southern part of the state. The entrance of Tertiary invaders probably began with the establishment of mixed hardwood forests in southern Michigan about 9,900 ybp and persisted through the warming period that lasted until about 2,500 ybp.

The Harper Holocene site near Lansing, Michigan, yielded four species of turtles, indicating that the modern herpetofauna of the area might have been in place by about 6,000 radiocarbon years ago. Additional evidence provided by Holman (1992b) indicates the possibility that herpetological species are still continuing to invade northern habitats in Michigan, especially where forest has been cleared and the area opened to more sunlight.

In summary, in the Great Lakes region, the reinvasion of herpetological species into areas covered by the Laurentide ice sheet was an important event. Unfortunately, Pleistocene herpetofaunal records are few in this region and are mainly confined to the late Wisconsinan. Nevertheless, the available evidence indicates that invading herpetological species followed rather closely the

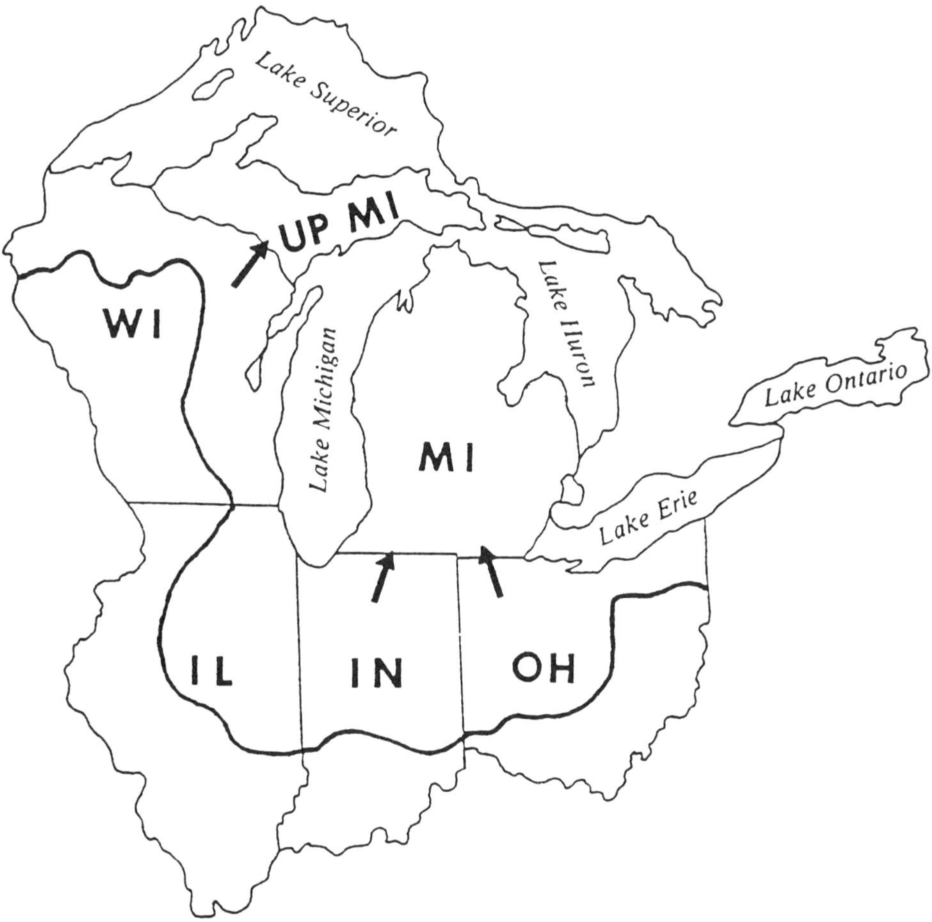

FIGURE 41. Map of the central Great Lakes region showing (1) major routes of herpetological reoccupation of postglacial Michigan and the approximate maximum extent (heavy line) of the Laurentide ice sheet in Ohio, Indiana, Illinois, and Wisconsin. [From Holman, 1992b]

retreating ice sheet at the end of the Wisconsinan. A model of this reinvasion based on geological, paleobotanical, vertebrate paleontological evidence, as well as ranges and ecological tolerances of modern herpetological species, suggests that herpetological species adapted to specific vegetational communities south of the glacial front invaded deglaciated areas in the order in which the reestablished vegetational communities arose. The possibility was presented that some herpetological species may still be moving northward where forests have been cut and the land exposed to more sunlight. Only one extinct species, the giant tortoise *Geochelone crassiscutata,* from the Sangamonian of central Illinois, has been recorded from the region.

Appalachian Region: Stability in the Midst of Change

Pleistocene herpetological species reinvaded vast areas of deglaciated land in the Great Lakes region. But well south of glacial penetration, in the Appalachian region, the herpetofauna was relatively stable. No extinct species have been reported from the region with the exception of a questionable extinct salamander (*Cryptobranchus guildayi*) and the large land tortoise *Geochelone*. A few extralimital species have been reported in both Irvingtonian and Rancholabrean localities, but these are not striking. On the other hand, mammalian faunas from the Pleistocene of the Appalachian region of North America contain many extinct taxa as well as striking northern and sometimes northwestern extralimital taxa.

The late Wisconsinan evidence of herpetological stability in the Appalachian region is particularly compelling because the data have been derived from many very diverse faunas, some of which are stratified and have been radiocarbon dated. The main extralimital species are a few southern ones (see Fay, 1988). The Ladds Quarry fauna near Cartersville in northwestern Georgia in the southermost part of the Appalachian region and the Cheek Bend Cave fauna from the Nashville Basin of southern central Tennessee, however, are noteworthy exceptions.

The Ladds fauna, which yielded thirty-three herpetological taxa, had the northern extralimital taxa *Clemmys insculpta, Elaphe vulpina,* and *Lampropeltis triangulum triangulum,* as well as the southern extralimital species *Bufo terrestris, Pseudemys nelsoni,* and *Geochelone crassiscutata*. The Cheek Bend fauna yielded the northern extralimital species *Clemmys insculpta* and the southern extralimital species *Bufo* cf. *Bufo terrestris* and *Rhadinaea flavilata*.

Why these herpetofaunas differ from other Appalachian herpetofaunas is not clear. Perhaps their lower latitudinal and altitudinal positions, and their better access to north–south trending river valleys, relative to the other Appalachian herpetofaunas, is an important factor.

The fact that Pleistocene herpetofaunas from the Appalachian Region usually have modern analogues although they usually co-occur with disharmonious mammalian faunas without modern analogues seems to be paradoxical. Several suggestions have been made in an attempt to explain this.

Unlike the modern mammalian fauna, the North America herpetofauna has no strictly tundra and boreal species that may be easily detected in Pleistocene faunas. In fact, most of the species that occur at high northern latitudes (e.g., *Rana sylvatica, Rana pipiens, Bufo americanus, Chelydra serpentina, Chrysemys picta, Thamnophis sirtalis*) occur in the Appalachian region today. Thus, if populations of these amphibians and reptiles were displaced southward into the Appalachian region by glacial activity in the Pleistocene, they would not be detected in the faunas. This seems like a very reasonable way to explain the apparent lack of "northern" intrusive species in the fauna.

It has also been suggested (Fay, 1988) that herpetological species have a marked capacity for acclimatization to seasonal climatic change, which may also provide the ability to adapt to progressive long-term changes by alterations in physiology and behavior; and that for this reason they did not respond as strongly to glacial–interglacial climatic changes as did other vertebrates.

A third suggestion, based on the study of New Trout Cave in West Virginia (Holman and Grady, 1987), is that the herpetofauna of the deposit might represent amphibians and reptiles from the vicinity of the cave, which is a quite restricted environment, but that mammalian fauna may have been transported by predators from wide ranging localities, thus representing a more regional aspect of the paleoenvironment. Thus, if a mosaic community existed, a wider variety of the mammalian species would be sampled, and northern and western intrusive mammalian species would be more likely to be taken.

In summary, numerous large Appalachian herpetofaunas, mainly late Wisconsinan, have been studied. Most of the herpetological species identified from these faunas are extant and living in or near the area today, quite unlike the mammalian faunas, which tend to have many extinct species and many northern and western extralimital forms. Explanations of this paradox are mainly inconclusive other than the fact that, in the east, no strictly boreal or tundra amphibians and reptile species exist today, so that southern movements of these taxa would not be detected in Pleistocene faunas. Only two extinct herpetological taxa, the questionable salamander *Cryptobranchus guildayi* and the large extinct tortoise, *Geochelone crassiscutata* have been identified from Pleistocene deposits in the Applachian region.

Southeastern Region: Much Stability, Some Extinction, and a Few Northern Invaders

More temporal units of the Pleistocene are represented in the Southeastern region than in the Appalachian region. In fact, in Florida alone, herpetofaunas bridge the gap from the earliest Irvingtonian through the latest Pleistocene and early Holocene.

A study of the many southeastern herpetofaunas reveals no indication of classic "glacial" and "interglacial" climates in this area so far removed from glacial activity. Nevertheless, there appears to have been more extinction of herpetological species than in the Appalachian region and several northern invading species have been identified. As an example of this, three large Florida faunas representing (1) the early Irvingtonian, (2) the Sangamonian, and (3) the late Wisconsinan will be discussed.

The single large Irvingtonian herpetofauna (IrI) in the Southeastern region is the Inglis IA fauna in southwestern peninsular Florida. This assemblage consists of at least forty-seven species and, not unexpectedly, has more extinct and extralimital taxa than most later Pleistocene faunas of the region. Extinct forms are *Geochelone crassiscutata, Geochelone incisa, Eumeces carri, Diadophis elinorae,* and *Regina intermedia*. Western extralimital forms include cf. *Gerrhonotus* and *Heterodon nasicus,* and northern extralimital forms include

Ambystoma sp., *Eumeces* cf. *Eumeces fasciatus,* and *Opheodrys vernalis.*

The most diverse Sangamonian herpetofauna of the Southeastern region is found at Reddick, in north-central Florida. This site yielded over fifty amphibian and reptile species. Extinct forms are *Geochelone crassiscutata,* and *Geochelone incisa* and the extinct subspecies *Trachemys scripta petrolei.* Northern invaders consisted of *Bufo woodhousii fowleri, Eumeces* cf. *Eumeces fasciatus,* and *Carphophis amoenus.*

In contrast to this, the large late Wisconsinan Vero fauna (at least thirty-four herpetological taxa) from southwestern peninsular Florida had a single extinct species, *Geochelone crassiscutata,* and only two extralimital northern species (*Ambystoma* sp. and *Sternotherus minor*), both found farther north in the peninsula today.

In turning to northern Mississippi, the very northern turtle *Emydoidea blandingii* occurred in the same late Pleistocene fauna with a very large specimen of the giant tortoise *Geochelone crassiscutata.* In general, occurrences of northern amphibians and reptiles with giant tortoises are thought to indicate a degree of climatic equability in the Southeastern region, as the giant tortoises are thought to exist only in areas that were frost free or essentially frost free (Holman, 1976). The relict occurrence of the mainly northern turtle *Clemmys guttata* in northern Florida today might lend additional credence to this hypothesis.

In summary, the herpetofauna of the Southeastern region of the United States remained relatively stable throughout the Pleistocene and indicated no evidence of classic glacial and interglacial climates. But in general, the Southeastern region showed more evidence of invasion by northern extralimital species and a higher rate of extinction of species than did the Appalachian region. The fact that intrusion of northern elements would be easier to detect in the Southeast than in Appalachia might be a factor.

Ozark Region: Stability and Some Questionable Southern Intruders

The isolated Ozark region, carved out of an ancient Paleozoic uplifted area, has yielded relatively few Pleistocene herpetofaunas. But these are very interesting, and some important unanswered questions about them exist. Ozark region herpetofaunas mainly have been found in caves, and caves are notoriously difficult to study because of stratigraphic problems discussed earlier in this book. Therefore, some of these caves may have a few early Holocene intrusive species mixed in with late Wisconsinan species.

Only one Irvingtonian herpetofauna, the Conard Fissure site in Newton County, Arkansas, has been detailed, and the rest represent late Wisconsinan times with some possible early Holocene intruders. Nevertheless, the indication is that the Ozark region, also entirely south of maximum glacial penetration, had a relatively stable herpetofauna from Irvingtonian times to the present.

The Irvingtonian II Conard Fissure site had at least fourteen species of amphibians present. None of these represent extinct forms, and only one extralimital northern species, *Elaphe vulpina,* was present. The most diverse Pleistocene site, the Crankshaft Cave late Wisconsinan–early Holocene site in Jefferson County near St. Louis, produced at least seventeen herpetological species but had no extinct or extralimital species present.

There are records of at least two worrisome species that do not fit this pattern. Two Missouri records of "? *Geochelone*" exist, one from Crevice Cave and another from Tom Moore Cave, both in Perry County. "*Alligator*" has been reported from Herculaneum Crevice in Jefferson County, Missouri, on the basis of osteodermal scutes, and it was reported that a Columbia, Missouri, drugstore had an alligator jaw in the window that was said to be from a local cave (Mehl, 1962). These alligator records need confirmation.

The fact that giant tortoises have been reported from Pennsylvania, Illinois, and northwestern Georgia does not make the Missouri records seem as doubtful. But rationalizing the presence of alligators in the Ozark region in the Pleistocene is difficult.

In summary, although Pleistocene herpetofaunas from the Ozark region are sparse and represent rather limited times in the Pleistocene, it appears that the herpetofauna of the region was relatively stable from Irvingtonian times to the present. Two questionable records of giant tortoises (*Geochelone*) and of *Alligator* need to be confirmed.

Northern Plains Region: General Stability, but Evidence of Extralimital Northern and Eastern Species

The Northern Plains region has been a critical area for Pleistocene herpetological studies. First, it lies to the west of the states that were deeply penetrated by the Wisconsinan Laurentide ice sheet; and second, herpetofaunas are available from more Pleistocene temporal units than in any other region of North America.

An important set of range adjustments that may be detected in the herpetofauna of the Northern Plains region consists of invasions of northern and eastern species during glacial ages, as well as other odd combinations of extralimital species. As in previously discussed regions, however, the Pleistocene herpetofauna of the Northern Plains region was quite stable during the Pleistocene compared to the avian and mammalian faunas. A few important herpetofaunas from this region are noted next.

The Java local fauna in Walworth County, in northern central South Dakota, contains the northernmost assemblage of Pleistocene amphibians and reptiles known from North America. It is considered to be of early Irvingtonian (IrI) age. There is no area where all the Java amphibians and reptiles could be found living together today. Forms that live south and southeast of the area are *Rana catesbeiana, Eumeces septentrionalis, Nerodia sipedon,* and *Elaphe vulpina. Phrynosoma douglasii* occurs today about 130 km west of Java.

The Albert Ahrens local fauna from Nuckolls County in southeastern Nebraska is considered to represent a full glacial period during Kansan times and is the most diverse Irvingtonian herpetofauna known from the Great Plains (Ford, 1992). Although all the herpetological taxa are living today, considerable range adjustments occurred in the fauna, as northern and northeastern (*Ambystoma laterale, Rana sylvatica, Elaphe vulpina*), eastern (*Bufo americana, Hyla chrysoscelis* or *versicolor, Rana clamitans, Thamnophis proximus, Crotalus horridus*), southern and southeastern (*Pseudacris* cf. *Pseudacris clarkii, Regina grahamii, Agkistrodon* sp.), and western (*Crotalus viridis*) forms occur in the fauna. The nearest area of sympatry of most of the Albert Ahrens species today lies in extreme northwestern Missouri. Ford (1992) suggested that this herpetological assemblage lived in a more equable climate than occurs in the area today.

The Courland Canal–Hall Ash assemblage from Jewell County in north-central Kansas is considered to represent early Kansan times (Rogers, 1982). Rogers found northwestern (*Bufo hemiophrys*) and southeastern elements (*Trachemys scripta, Agkistrodon contortrix*) in this fauna and concluded that a more equable climate may have occurred in Kansas at this time.

The Williams local fauna of Rice County in central Kansas is considered to represent Illinoian times. The presence of extralimital forms (*Emydoidea blandingii,* and *Elaphe vulpina*) that occur mainly northeast of the area today is thought to have indicated somewhat moister, cooler conditions. But a full glacial climate or a dominant coniferous forest community is not indicated by the rest of the herpetological species, all of which can be found in the area today (Holman, 1984).

The Mount Scott local fauna of Meade County in southwestern Kansas is thought to represent the Illinoian glacial age based on the faunal assemblage (Holman, 1987a). Northern (*Emydoidea blandingii, Elaphe vulpina*) and eastern (*Sternotherus odoratus, Terrapene carolina* ssp., *Ophisaurus ventralis, Regina grahamii*) forms place the only modern area of sympatry for the Mount Scott local fauna herpetofauna in extreme northeastern Missouri. The Mount Scott fauna also contains the extinct turtle *Pseudemys hibbardi.*

The Butler Spring locality (including the so-called sloth locality) of Meade County in southwestern Kansas yielded a large herpetofauna that is considered to represent the Illinoian age. Several members of the Butler Spring fauna are extralimital, living to the north (*Emydoidea blandingii, Elaphe vulpina*) and east (*Pseudemys* cf. *Pseudemys concinna hieroglyphica, Ophisaurus* cf. *Ophisaurus attenuatus, Regina grahamii, Storeria* cf. *Storeria dekay*) of the area today. In fact, this herpetofauna would be typical of the Illinoian rather than the Kansan biotic province in which Meade County, Kansas, lies today.

The Cragin Quarry fauna, also in Meade County, yielded a large herpetofauna that is considered to be late Sangamonian in age. This is the type locality of the extinct toad *Bufo repentinus* and the extinct medium to large-sized tortoise

Geochelone equicomes. This fauna contained extralimital eastern (*Pseudacris* cf. *Pseudacris triseriata, Terrapene carolina* ssp., *Agkistrodon contortrix*), southern (*Phrynosoma modestum, Cophosaurus* cf. *Cophosaurus texanus*), and possibly southern (an old record of "*Alligator*") forms. This would indicate warmer, moister Sangamonian times.

The Jinglebob local fauna of Meade County also yielded a large herpetofauna and is considered to represent pre–radiocarbon-datable Wisconsinan times. Several members of the Jinglebob fauna are extralimital forms that occur to the north (*Pseudacris* cf. *Pseudacris triseriata, Emydoidea blandingii*) or east (*Terrapene carolina* ssp., *Ophisaurus attenuatus, Elaphe obsoleta*) of Meade County, Kansas, today.

Finally, the Robert Fauna of Meade County yielded a very late Wisconsinan fauna that existed about 11,000 radiocarbon ybp. The Robert fauna yielded ten species of fossil snakes that occur in Meade County, Kansas, today, with the exception of *Regina grahamii* and *Storeria* cf. *Storeria dekayi* which occur somewhat east of Meade County.

In summary, diverse herpetofaunas representing several units of the Pleistocene occur in the Northern Plains region. These range from early Irvingtonian to late Wisconsinan in age. Extralimital species have been recorded from many sites in this region. Several Illinoian faunas contained northern and eastern extralimital species, while other faunas of varying ages had eastern, western, and even southern extralimital intruders. These faunas tend to illustrate the previously discussed concept of the individuality of species responses to Pleistocene climatic changes, and the Pleistocene equability hypothesis may be invoked to explain the presence of extralimital northern, eastern, southern, and/or southwestern species occurring together in several Northern Plains region faunas. An extinct toad species, *Bufo repentinus,* and the extinct turtle species *Pseudemys hibbardi* occur in this region. Moreover, specimens of extinct tortoises of the genera *Geochelone* (including the type specimen of *Geochelone equicomes*) and *Gopherus* are found in several faunas in the region.

Southern Plains Region: General Stability, but Extinction of Several Turtle Taxa

The Southern Plains region is well represented by Pleistocene herpetofaunas, but these are mainly of late Wisconsinan age. This is quite different from the situation in the Northern Plains region, where late Wisconsinan herpetofaunas are rare. The picture that is presented by the herpetofaunas of the Southern Plains region is one of relative stability except for the extinction of several taxa of turtles. In general, there were relatively few extralimital species present in these faunas.

The most important extinct and/or extralimital taxa are listed below. The northern turtle *Emydoidea blandingii* occurs in the late Irvingtonian Berends local fauna of Beaver County, Oklahoma, and Doby Springs local fauna of

Harper County, Oklahoma. It also has been identified in the Rancholabrean Nye Sink locality of Beaver County, Oklahoma. The nearest this turtle gets to Oklahoma today is southern central Nebraska (Conant and Collins, 1991).

An extinct species of aquatic turtle, *Pseudemys hibbardi* (Fig. 36), was described from the late Irvingtonian Doby Springs fauna of Harper County, Oklahoma. An extinct subspecies of slider turtle, *Pseudemys scripta bisornata,* has been reported from the early Irvingtonian of the Gilliland fauna of Knox County, Texas.

A large, extinct *Geochelone* sp. ("giant land tortoise") has been found in the Irvingtonian Gilliland fauna of Knox County, Texas and the Fyllan Cave fauna of Travis County, Texas. Large *Geochelone* have also been found in the following Rancholabrean sites: Easley Ranch local fauna of Foard County, Texas; Clear Creek fauna of Denton County, Texas; Nye Sink locality of Beaver County, Oklahoma; Pitt Bridge localities of Brazos and Burleson counties, Texas; Lewisville site of of Denton County, Texas; Moore Pit Local fauna of Dallas County, Texas; Sims Bayou fauna of Harris County, Texas; and the Ingleside fauna of San Patricio County, Texas.

A small extinct tortoise species, *Geochelone wilsoni* (Fig. 43C), has been found in Rancholabrean sites in Oklahoma and Texas. These include the Shulze Cave site in Edwards County, Texas; the Friesenhahn Cave site in Bexar County, Texas, and the Domebo locality in Caddo County, Oklahoma. A small, unnamed extinct species of *Geochelone* has been found in Irvingtonian faunas including the Gilliland local fauna of Knox County, Texas, and the Berends local fauna of Beaver County, Oklahoma.

Extinct large burrowing tortoises of the genus *Gopherus* have been found in Rancholabrean faunas. These include the Easley Ranch fauna of Foard County, Texas; Nye Sink locality of Berends County, Oklahoma; and the Ingleside fauna of San Patricio County, Texas. These probably all represent *Gopherus laticauda.* An extralimital *Gopherus,* resembling the extant species *Gopherus polyphemus,* was identified from the Irvingtonian Gilliland fauna of Knox County, Texas, far northwest of its occurrence in the southeastern United States today.

A large form of the modern box turtle *Terrapene carolina* has been found in Rancholabrean sites over a large part of Texas reaching west of its modern range in Lubbock County. These turtles are thought to be genetically related to the large, extinct subspecies of box turtle, *Terrapene carolina putnami,* that occurs in "pure form" in Florida.

Alligators have been found north and west of their range today in Texas at the early Irvingtonian Gilliland fauna of Knox County, Texas, and west of their modern Texas range in the Slaton local fauna of Lubbock County, Texas.

In summary, diverse Pleistocene herpetofaunas are abundant in the Southern Plains region, especially from late Wisconsinan sites in Texas. None of the herpetological taxa reported from the Southern Plains region became extinct during the Pleistocene except for several turtles, including *Pseudemys hibbardi, Trachemys scripta bisornata, Geochelone* large sp. indet., *Geochelone wilsoni,*

Geochelone laticauda, and a very large form of the modern box turtle species *Terrapene carolina* that supposedly is genetically related to the extinct subspecies *Terrapene carolina putnami.*

Extralimital modern herpetological species are not especially numerous in the Southern Plains region, the most important ones being *Emydoidea blandingii,* a northern species that had extended south into the region during the Pleistocene, and *Alligator mississippiensis,* a Pleistocene form which ranged north and west of its modern range in Texas today.

Rocky Mountain/Great Basin Region: Adjustments in the Arid Southwest

The vast, mountainous, and arid Rocky Mountain/Great Basin region has produced many interesting Pleistocene herpetofaunas, several of which have been very well-studied and have had stratigraphic control. Many of these faunas have been associated with packrat middens in caves or rock shelters. These herpetofaunas contain no extinct species with the exception of an undescribed species of *Ambystoma* from an Irvingtonian site in Colorado and the extinct small tortoise *Geochelone wilsoni* from two Rancholabrean sites. Two striking extralimital forms are *Chelydra serpentina,* which existed far west of its modern range in the Pleistocene, and *Gopherus agassizi,* which existed east of its modern range.

One of the important outgrowths of the study of Pleistocene amphibians and reptiles in the Rocky Mountain/Great Basin region is the elimination of the scenario that the desert herpetofauna of the American Southwest retreated to Mexican refugia during glacial periods, as many desert species were found in their original ranges during the late Wisconsinan glacial age.

Some very important studies in the region have dealt with radiocarbon dated changes in herpetofaunas through time in the same area or site. In some of these studies it has been shown that herpetofaunas tend to have remained stable even though the vegetational associations around them changed during the late Pleistocene. These studies have also shown that above and beyond general taxonomic stability, few temporal changes in faunal composition occur. Some important records and sites are summarized below.

Remains of the snapping turtle *Chelydra serpentina* were found in the Glendale fauna in Clark County, Nevada (Van Devender and Tessman, 1975). The fossils occur 40 cm above a radiocarbon-dated charcoal that yielded a radiocarbon date of greater than 33,000 ybp. The nearest living populations of this species occur in streams along the eastern part of the continental divide at least 650 km away.

The Hansen Bluff Irvingtonian site (IrII; Repenning, 1987) consisted of superimposed layers of sediments 2,300 m high that could be correlated with ash deposits and reversed and normal magnetic polarities (Rogers et al., 1985). Here, based on palynological and faunal analysis, warm-dry periods alternated with cold-wet ones. Warm-dry periods existed from 840,000 to 823,000 ybp and from 790,000 to 773,000 ybp with a cold-wet period between them. During the warm-dry periods combinations of *Scaphiopus bombifrons, Bufo* sp., *Rana catesbeiana, Holbrookia maculata,* and *Heterodon nasicus* occurred in the sediments; during the cold-wet period between them only *Rana* sp. was found. A cold-wet period that occurred between 760,000 and 740,000 ybp had no herpetological species.

The Dry Cave site near the Texas border in Eddy County in southeastern New Mexico lies in the ecotone between woodland and Chihuahuan desert vegetation. The site represents a time period of between 25,000 and 35,000 ybp. Here the desert tortoise, *Gopherus agassizi,* was found well east of its present range in the Sonoran and Mojave deserts. The extinct tortoise *Geochelone wilsoni* also occurred at this site. These animals, along with other faunal elements at the site, indicate that Dry Cave was subjected to no more than light freezing during the time that the fauna accumulated (Harris, 1987).

A series of herpetofaunas from packrat middens in the lower Grand Canyon (Rampart Cave, Vulture Canyon, Emery Falls Canyon; Rocky Mountain/Great Basin section, Chapter 6, this book) reported by Van Devender et al. (1977) ranged in age from 8,500 to 16,330 ybp. The composite herpetofauna represented a mixture of modern desert species such as *Gopherus agassizii, Sauromalus obesus,* and *Coleonyx variegatus* and higher woodland species such as *Sceloporus undulatus* or *occidentalis.* Along with other faunal elements, it was determined that these species indicated a more moderate climate with mild winters, cool summers, and somewhat increased winter rainfall.

A series of herpetofaunas from packrat middens ranging in age from Pleistocene to Holocene in Arizona (Wolcott Peak, Brass Cap Point, New Water Mountains, and Welton Hills; Rocky Mountain/Great Basin section, Chapter 6, this book) reported by Van Devender and Mead ranged in age from 14,500 to 5,000 ybp. These herpetofaunas were associated with juniper or pinyon-juniper woodlands. It was shown that the composite herpetofauna has been relatively stable for about the last 15,000 years, while the vegetation has changed dramatically!

Another important outgrowth of this study was that biogeographical hypotheses calling for restriction of desert herpetological species to Mexican refugia during glacial periods were negated, as many desert species were present in the late Wisconsinan glacial segments of the faunas.

Moreover, equable Pleistocene climates with mild winters, cool summers, and increased precipitation are inferred, based in part on the herpetofauna. Species from the Arizona sites wholly or mainly restricted to desert scrub habitats today include *Gopherus agassizii, Coleonyx variegatus, Sauromalus obesus, Uta stansburiana,* and *Chionactis occipitalis.* A species that is common in higher, more mesic habitats today is *Sonora semiannulata.* The climatic implications of animals restricted to desert scrub habitats today occurring in late Pleistocene woodlands, as is the case with those of the faunas listed above, is that the winters were probably no colder than today.

A fossil herpetofauna from Smith Creek Cave in White Pine County in the Snake Range of Nevada yielded a herpetofauna from packrat middens that ranged from 13,000 to 12,000 ybp (Mead et al., 1982). Of much importance is the fact that the following Great Basin desert reptiles occur in this late Wisconsinan glacial time: *Gambelia wislizenii, Phrynosoma platyrhinos, Sceloporus magister, Hypsiglena torquata,* and *Rhinocheilus lecontei.* This negates the hypothesis of Tanner (1978) that most of the present Great Basin desert reptiles have extended their ranges into the Great Basin as post-Pleistocene introductions from Pleistocene refugia in Mexico.

The Brown Sand Wedge fauna from Roosevelt County in New Mexico represents a site that was deposited at about 11,000 ybp. This site contained the only records of the eastern box turtle, *Terrapene carolina* ssp., and the western form, *Terrapene ornata* from the Rocky Moutain/Great Basin area of this book. The western box turtle occurs in New Mexico today, but the eastern box turtle ranges no farther west than eastern Texas in the modern fauna. The small extinct Pleistocene tortoise *Geochelone wilsoni* also occurs in the Brown Sand Wedge fauna.

At Crystal Ball Cave in Millard County, Utah, fossils from packrat middens and storm runoff yielded herpetofaunal remains that ranged from about 23,000 ybp to late Holocene in age (Mead et al., 1989). Here *Phrynosoma douglasii,* a form of relatively high elevations in the area today, occurred with forms typical of lower elevations.

In summary, the herpetofaunas of the Rocky Mountain/Great Basin area remained generally very stable from Pleistocene into Holocene times, with extinction of only three species, an unnamed salamander *Ambystoma, Geochelone wilsoni,* and a snake, *Lampropeltis intermedius* (Fig. 38) of the Irvingtonian of Arizona. Extralimital western occurrences of *Chelydra serpentina* in Nevada and *Terrapene carolina* ssp. in New Mexico are striking, as is the eastern occurrence of the desert tortoise *Gopherus agassizi* in New Mexico.

Two important ecological and zoogeographical considerations that grew out of studies of Pleistocene herpetofaunas in the Great Basin are (1) that Pleistocene reptile species living in desert areas today were able to adapt to woodland situations (e.g., pinyon-juniper associations) during the Pleistocene when equable climatic conditions existed and (2) that desert reptiles in the Great Basin area did not retreat to Pleistocene refugia in Mexico during the late Wisconsinan.

Pacific Coast Region: Stability and Adjustment

The herpetofaunas of the Pacific Coast region are relatively few compared to many other regions of this book. These faunas have come from a variety of situations ranging from tar pits to caves. The herpetofauna of this region has remained relatively stable from about 40,000 years ago to the present. Only one species of rattlesnake, the questionable *Crotalus potterensis* (Fig. 39), and a tortoise, (?) *Geochelone,* are extinct forms.

Extralimital forms are not abundant, but, as in the Rocky Mountain/Great Basin region, some adjustments to vegetational changes were made. Pleistocene herpetofaunas from the Pacific Coast region also strongly negate the hypothesis that desert reptiles retreated to Mexican refugia in the Pleistocene. Some important faunas that deal with these subjects are discussed below.

The Costeau Pit of Orange County, California, has yielded a fauna which is believed to be about 40,000 years old (Hudson and Brattstrom, 1977). All the animals are extant and all occur in the area today with the exception of *Gopherus,* which occurs in desert areas to the east of the site today.

The Rancho La Brea site in Los Angeles County, California, has yielded the largest herpetofauna of the Pacific Coast region. It is a fluvial site where entrapment occurred in asphalt-impregnated streams and ponds. At least twenty-eight species of amphibians and reptiles were identified from this site of which only one, "? *Geochelone* sp.," is extinct, if indeed *Geochelone* is actually represented. An extralimital lizard, *Sceloporus magister,* occurs today in more xerophytic areas to the east.

The Potter Creek Cave site in Shasta County, California, yielded rattlesnake remains, which were described as an extinct species, the questionable *Crotalus potterensis* (Brattstrom, 1953a).

A series of herpetofaunas from Sonoran Desert packrat middens in California (Van Devender and Mead, 1978) ranging from Pleistocene to earliest Holocene (Falling Arches, Redtail Peaks, Tunnel Ridge, and Whipple Mountain; Pacific Coast section, Chapter 6, this book) ranged in age from 14,000 to 9,000 ybp. These herpetofaunas were associated with juniper and pinyon-juniper woodlands. It was shown that the composition of the herpetofauna has been relatively stable for the past 14,000 years, while the vegetation has changed remarkably. Another outgrowth of this study was that biological scenarios postulating that United States desert herpetological species were restricted to Mexican refugia during glacial periods were negated by the presence of desert species during the late Wisconsinan.

Moreover, equable Pleistocene climates with mild winters and cool summers and increased precipitation were inferred. Species from these Pleistocene herpetofaunas that are "obligate" associates of desert scrub today are *Gambelia wislizenii, Sauromalus obesus, Lichanura trivirgata,* and *Chionactis occipitalis. Lampropeltis getula* is a common species of higher and more mesic situations. The climatic implication of animals restricted to desert scrub habitats today occurring in Pleistocene woodlands, as is the case with those of faunas listed above, is that winters were probably no colder than today.

In summary, the herpetofauna of the Pacific Coast region was relatively stable from about 40,000 years ago to the present. Only two species, a rattlesnake, the questionable *Crotalus potterensis,* and a possible tortoise, (?) *Geochelone* sp., became extinct. Extralimital species were not abundant, but evidence indicates that many herpetological species that occur in desert situations today were able to exist in woodland habitats in the Pleistocene when more equable climates existed. The scenario that United States desert species retreated to Mexican refugia during Pleistocene glacial periods is refuted by the presence of desert species in late Wisconsinan faunas.

8

EXTINCTION PATTERNS IN THE NORTH AMERICAN PLEISTOCENE HERPETOFAUANA

The "why" of the extinction of mammalian families, genera, and species—and to a somewhat lesser extent, bird taxa—during and at the end of the Pleistocene has been a hotly debated issue in recent decades (e.g., Martin and Wright, 1967; Kurten and Anderson, 1980; Martin and Klein, 1984). The general impression among paleoherpetologists, however, is that the continental North American herpetofauna has been in a sort of "holding pattern" for the past 1.9 million years (Holman, 1991b).

The truth of the matter is that many paleoherpetologists choose not to study Pleistocene amphibians and reptiles because they think there is not much exciting to learn. On the contrary, it would seem that if this herpetological stability in the face of Pleistocene changes is really true, the question "What attributes have allowed the herpetofauna to survive the stresses that have caused dramatic extinctions in other classes during the Pleistocene?" is an extremely provocative one.

We shall first consider extinction patterns in continental North America (excluding Mexico) based on an analysis of the 170 faunas detailed in this book thus far. This will include a short discussion of the validity of some of the Pleistocene herpetological taxa that are currently considered to be extinct. Then we shall discuss some possible modifications of what we call the "Pleistocene herpetological stability concept," and finally we will address "whys" of this proposed stability and possible future areas of study.

Late Pliocene (Blancan) Herpetofaunas

Late Blancan herpetofaunas near the onset of the Pleistocene are rare; only two important sites are known from this narrow period of time. These are the Big Springs Quarry fauna of Antelope County in northeastern Nebraska and the Hornet's Nest fauna of Knox County in northeastern Nebraska (Rogers, 1984).

The Big Springs site dates at about 2.5 million ybp (Voorhies, 1987). The Hornet's Nest fauna dates at about 2 million ybp (Holman and Schloeder, 1991), just before the onset of the Pleistocene.

Thirty-one herpetological taxa were identified in the Big Springs Site with five species (16.1 percent) being reported as extinct. At the Hornet's Nest site, eighteen herpetological taxa were identified , with only one species (5.6 percent) being extinct. This apparent shift from 16 percent to about 7 percent extinction over a half-million year period is probably somewhat of an artifact as the 3.8 million ybp Lisco C fauna of Garden County in western Nebraska (Holman and Schloeder, 1991) and the 3.5 million ybp Sand Draw fauna of Brown County in north-central Nebraska both had low herpetological extinction percentages: 6 percent of eighteen taxa in the Lisco C fauna, and 11 percent of nineteen taxa in the Sand Draw fauna respectively.

The misleading difference between the 6 percent rate of extinction in the much older Lisco C herpetofauna and the 16 percent rate in the younger Big Springs herpetofauna relates to the fact that three *Bufo* were identified as extinct species in the Big Springs (Rogers, 1984), whereas only *Bufo* sp. was identified in the Lisco C. In any case, it appears that the herpetofauna of the Blancan of Nebraska must have been essentially modern at the onset of the Pleistocene.

A Look at "Extinct" North American Pleistocene Herpetological Taxa

Several "extinct" North American herpetological taxa that were described by paleontologists in the early twentieth century have long been synonymized with living forms and will not be discussed here. Contemporary extinct herpetological taxa that appear in Pleistocene faunal summaries fit into two categories: (1) those that have not been questioned by modern workers, and (2) those that have been rather recently questioned. We shall discuss some of the more important taxa in each of these categories. Following is a list of currently recognized extinct North American (exclusive of Mexico) herpetological taxa:

Cryptobranchidae
- *Cryptobranchus guildayi* Holman, 1977a: Irvingtonian of Maryland and West Virginia (Figs. 16, 30)

Sirenidae
- *Pseudobranchus robustus,* Goin and Auffenberg, 1955: Rancholabrean of Florida (Fig. 33)

Bufonidae
- *Bufo repentinus* Tihen, 1962: Rancholabrean of Kansas (Fig. 34)
- *Bufo woodhousii bexarensis* Mecham, 1959: Rancholabrean of Texas (Fig. 37)

Hylidae
- *Hyla baderi* Lynch, 1965: Rancholabrean of Florida (Fig. 32)

Hyla holmani Lynch, 1966: Rancholabrean of Texas (Fig. 35)

Emydidae

Pseudemys hibbardi (Preston, 1979): Late Irvingtonian to Rancholabrean of Oklahoma and Kansas (Fig. 36)

Terrapene carolina putnami (Hay, 1907): see discussion to follow (Fig. 42A)

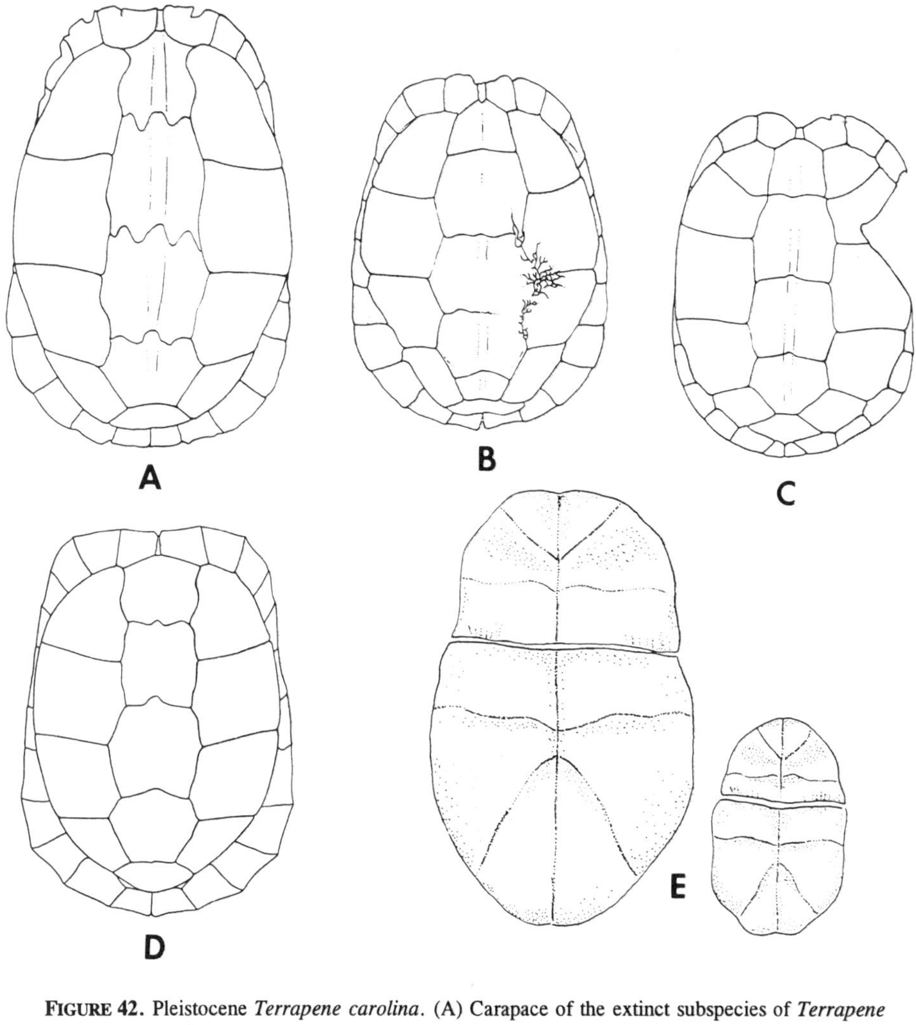

FIGURE 42. Pleistocene *Terrapene carolina*. (A) Carapace of the extinct subspecies of *Terrapene carolina putnami* from the Rancholabrean of the Melbourne site, Brevard County, Florida. (B) Modern *Terrapene carolina bauri* carapace from Alachua County, Florida. (C) Fossil *Terrapene carolina carolina* carapace from the Arredondo I Rancholabrean site, Alachua County, Florida. (D) Modern *Terrapene carolina major* shell from Gulf County, Florida. (E) Plastron of *Terrapene carolina putnami* (left) from the Rancholabrean Lewisville site, Denton County, Texas, compared with that of an adult male *Terrapene carolina carolina* (right) from Macoupin County, Illinois. [A–D from Auffenberg, 1958; E from Holman, 1965c]

Trachemys scripta bisornata (Cope, 1878): Irvingtonian of Texas
Trachemys scripta petrolei (Leidy, 1868): Rancholabrean of Florida

Testudinidae
Geochelone crassiscutata (Leidy, 1889): Rancholabrean of a large portion of the eastern United States (Fig. 43A)
Geochelone equicomes (Hay, 1917): Rancholabrean of Kansas
Geochelone incisa (Hay, 1916): Rancholabrean of Florida (Fig. 43B)
Geochelone mlynarski Auffenberg, 1988: Irvingtonian of Florida
Geochelone wilsoni (Milstead, 1956): Late Rancholabrean of Oklahoma, Texas, and New Mexico (Fig. 43C)
Gopherus laticaudus (Cope, 1893): Irvingtonian of Kansas and Texas

Scincidae
Eumeces carri Meylan, 1982: Early Irvingtonian of Florida

Colubridae
Diadophis elinorae Auffenberg, 1963: Early Irvingtonian of Florida (Fig. 31)
Lampropeltis intermedius Brattstrom, 1955a: Early Irvingtonian of Arizona (Fig. 38)
Regina intermedia Meylan, 1982: Early Irvingtonian of Florida

Viperidae
Crotalus potterensis Brattstrom, 1953a: Rancholabrean of California (Fig. 39)

No extinct families or genera of Pleistocene amphibians and reptiles are currently recognized in the North American continental Pleistocene (Holman, 1991b). Nevertheless, several extinct species still appear in current lists and some of them warrant discussion.

Cryptobranchus guildayi Holman (1977a), an extinct species of cryptobranchid salamander, was described from the Irvingtonian Cumberland Cave, Maryland (Figs. 16, 30). Following Estes, Naylor (1981) stated that this extinct species was probably referable to the living species *Cryptobranchus alleganiensis*. Estes (1981) had stated that characters used to define *Cryptobranchus guildayi* are variable in all modern specimens of *Cryptobranchus alleganiensis* available to him. Therefore, *Cryptobranchus guildayi* should probably be considered a *species inquirendae* until more fossil and recent material has been studied.

Pseudobranchus robustus Goin and Auffenberg (1955), an extinct species of sirenid salamander recorded from Rancholabrean localities in Florida (Fig. 33), was distinguished from the modern species *Pseudobranchus striatus* mainly based on the wider angle between the aliform processes in the fossil. But Estes (1981) stated that it is possible that further study will show that *Pseudobranchus robustus* belongs to one of the living species. Moreover, Lynch (1965) demonstrated that the angle between the aliform processes of *Pseudobranchus striatus* available to him (at that time) overlapped those of *Pseudobranchus*

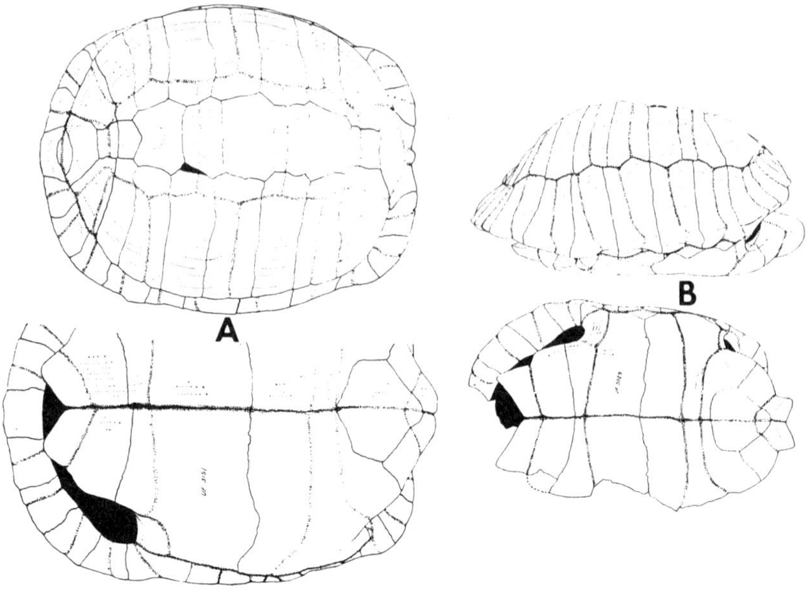

FIGURE 43. Extinct Pleistocene tortoises of the genus *Geochelone*. (A) *Geochelone crassiscutata* Leidy, a very large species (upper, carapace; lower, plastron). This specimen is from the Rancholabrean Haile VIII8 site, Alachua County, Florida. (B) *Geochelone incisa* Hay, a small, relatively thick-shelled species (upper, carapace; lower, plastron). This specimen is from the same

robustus. *Pseudobranchus robustus* then also fits the category of *species inquirendae*.

The extinct subspecies *Bufo woodhousii bexarensis* Mecham (1959) was described on the basis of a fragmentary femur and a few other elements from the late Wisconsinan of the Friesenhahn Cave in Texas (Fig. 37). Further studies are needed to confirm *Bufo woodhousii bexarensis* as a valid subspecies, as a fragmentary femur is a questionable element upon which to base a new taxon.

Hyla holmani Lynch (1966) was described on the basis of a fragmentary ilium from the Rancholabrean Howard Ranch (Groesbeck Creek) local fauna of Hardeman County, Texas (Fig. 35). Holman (1969b) reported that this ilium resembled that of a young *Hyla versicolor* so closely that *Hyla holmani* should be tentatively referred to that species until more Pleistocene and modern comparative material becomes available. It is the opinion of the author that this situation still exists, and that *Hyla holmani* should be considered a *species inquirendae*.

An extinct aquatic turtle species, *Pseudemys hibbardi* Preston (1979), from late Irvingtonian and early Rancholabrean sites in Oklahoma and Kansas, was described on the basis of the anterior part of the plastron (Fig. 36) and some

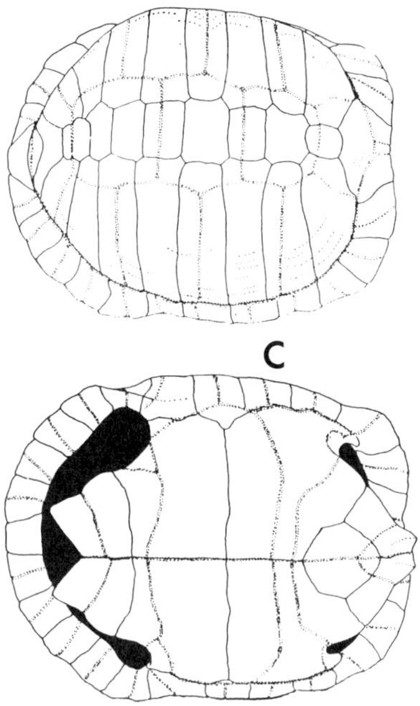

site as A. (C) Holotype (Texas Memorial Museum Number 933-3585) of *Geochelone wilsoni* Milstead (upper, carapace; lower, plastron) from the Rancholabrean Friesenhahn Cave site, Bexar County, Texas. This species is thought to be closely related to *Geochelone incisa*. [From Auffenberg, 1963a]

isolated pleural and peripheral bones. It was pointed out by Holman (1991b) that this form is quite similar to the modern species *Pseudemys concinna* and *P. floridana,* species that are usually distinguished from one another and related fossil species on the basis of carapacial rather than plastral characters. Until more is known about the variation in the anterior parts of the plastra of these species, or in the hybrids that are known to occur between these species, it was suggested that *Pseudemys hibbardi* be considered a *species inquirendae*.

Irvingtonian *Trachemys scripta* fossils from Texas have been identified as the extinct subspecies *Trachemys scripta bisornata* (Cope, 1878). Preston (1979) and Weaver and Robertson (1967) have suggested that all Rancholabrean slider turtles be designated as *Trachemys scripta petrolei*. Characters used to distinguish these subspecies appear to be individually variable or related to size differences (e.g., Preston, 1966; Weaver and Robertson, 1967; Preston, 1979). Thus, the use of these subspecific names appears to be tenuous and not particularly useful in an evolutionary sense.

A consideration of the extinct subspecies *Terrapene carolina putnami* (Hay, 1908) (Fig. 42A) is very complex as it involves the supposition of widespread gene exchange between the extinct form and three modern subspecies of box turtles, *Terrapene carolina bauri* (Fig. 42B), *T. carolina carolina* (Fig. 42C), and *T. carolina triunguis*. In fact, if this assumption of gene exchange is correct, Pleistocene box turtles of the *Terrapene carolina* ssp. complex document the intricacies and complexities of the speciation process over a geologically short period of time as much as any fossils have ever done.

Terrapene carolina putnami, as conceived in pure form, is a very large to "huge" (Fig. 42E) box turtle of over 200 mm in shell length that occurred in the Rancholabrean and became extinct at the end of the Pleistocene. Its very large size has been used in combination with other somewhat variable characters to distinguish it from the modern subspecies of *T. carolina.* This large species *Terrapene carolina putnami* is thought to have exchanged genes with *Terrapene carolina triunguis* in the west, with *Terrapene carolina carolina* to the northeast, and with *Terrapene carolina bauri* in peninsular Florida; many of these "hybrids" have been reported in various combinations in the fossil record (e.g., Auffenberg, 1958; Milstead, 1967; Preston, 1979) as well as "pure" *Terrapene carolina putnami.*

The modern subspecies *Terrapene carolina major,* a large subspecies (Fig. 42D) of the Gulf Coast area that exists in pure form only in northwest Florida (Conant and Collins, 1991), supposedly was formed by the interbreeding of *Terrapene carolina putnami* and *Terrapene carolina* subspecies in the Pleistocene. For some reason the very large box turtles (*Terrapene carolina putnami* and hybrids) became extinct at the end of the Pleistocene, leaving *Terrapene carolina major* as the only modern box turtle with a significant portion of *Terrapene carolina putnami* genes.

The most important extinct Pleistocene species in North America are the land tortoise of the genus *Geochelone* and to a lesser extent *Gopherus.* Extinct *Geochelone* land tortoises occurred from Pennsylvania and central Illinois southward through Florida, Georgia, Mississippi, and westward through Texas to New Mexico, perhaps reaching California. I have been conservative in the use of specific names following Preston (1979).

Neverthless, at leat six species became extinct at the end of the Pleistocene. The giant form *Geochelone crassiscutata* (Fig. 43A) was a very widespread form that ranged from Florida and Texas north to Pennsylvania and central Illinois in the eastern United States. A medium-sized to large form, *Geochelone equicomes,* occurred in Kansas and probably other midcontinental states.

Three small forms of *Geochelone* are recognized. *Geochelone mylnarskii* occurred in the Irvingtonian of Florida. *Geochelone incisa* (Fig. 43B) occurred in the Rancholabrean of Florida. *Geochelone wilsoni* (Fig. 43C) occurred in Texas westward into New Mexico. A large extinct form of the gopher tortoise, *Geochelone laticaudus,* is recognized in the Irvingtonian of Texas and probably occurred upward through Kansas as well. It might be suggested that all these species became extinct due to a combination of the onset of less equable climates and the invasion of human predators late in the Pleistocene.

Beginning with Hibbard (1960), it has been suggested that the large extinct tortoises of the genus *Geochelone* are indicative of warm, frost-free winters in the areas where they occurred in the Pleistocene. This is based on the fact that modern large *Geochelone* are not able to survive in areas where there are extended periods of cool weather.

Crotalus potterensis Brattstrom (1953a) (Fig. 39), an extinct fossil rattlesnake, was described from the Rancholabrean Potter Creek site of Shasta County, California, on the basis of vertebrae. Holman (1981) could find no characters on the vertebrae of *Crotalus potterensis* that distinguished them from from those of the modern species *Crotalus viridis*. Moreover, measurements reflecting size in the fossil vertebrae and in modern *Crotalus viridis* did not indicate the fossil form exceeded the modern form in size. Therefore, Holman (1991b) proposed that *Crotalus potterensis* be relegated to the synonymy of *Crotalus viridis*.

Comparative Extinction Patterns

I have attempted in this book to list all amphibian and reptile species recorded in all published Pleistocene herpetofaunas in continental North America (exclusive of Mexico) with the exception of a few scattered individual records that have been duplicated elsewhere in larger contemporaneous faunas. One hundred and seventy have been documented. Differences in extinction patterns between Pleistocene amphibians and reptiles and those of birds and mammals are striking.

In the late Pleistocene alone, 2 families and 19 genera of North American birds became extinct, with a very large number of extinct species reported (Holman, 1991b). Moreover, 8 families, 46 genera, and 191 species of mammals became extinct in the North American Pleistocene (Holman, 1991b). No families and no genera of amphibians and reptiles became extinct in the Pleistocene of North America, and only 12 taxa out of a total of 229 identified Pleistocene taxa were **unquestioned** extinct species.

Salamanders had no unquestioned extinct species among twenty-three identified taxa. Anurans had only two unquestioned extinct species out of forty-four identified taxa. Testudines, conservatively, had six unquestioned extinct taxa out of thirty-three identified taxa. These are all extinct tortoises of the genera *Geochelone* and *Gopherus*.

Lizards had only one unquestioned extinct species among 46 identified taxa and snakes had only three unquestioned extinct species out of seventy-six identified taxa. Amphisbaenians and crocodilians had no extinct species, but only one taxon each of these groups was identified in the North American Pleistocene.

Possible Modifications of the Pleistocene Herpetological Stability Concept

Before one speculates too much about the "why" of the apparent evolutionary stasis in the herpetofauna of the Pleistocene of North America, some possible modifications of the concept should be considered. As we have seen (Chapter 5), the great majority of identifications of Pleistocene amphibians and reptiles are made on the basis of certain individual skeletal elements that are considered to be diagnostic and that are frequently found as fossils.

Some of the more important of these elements are vertebrae in salamanders, ilia in anurans, portions of shells in testudines, dentaries in lizards, and vertebrae in amphisbaenians and snakes. Thus, one of the most serious problems is that extinct species and possibly even genera have gone unrecognized. Cryptic species such as the gray treefrogs *Hyla versicolor* and *Hyla chrysoscelis* have identical ilia as far as I can determine, and cryptic forms undoubtedly occurred in the Pleistocene.

Also problematic is the fact that many cases obtain where modern herpetological species can be readily identified on the basis of external morphology but are difficult or impossible to separate osteologically. Prime examples are frogs of the *Rana pipiens* complex of species; species of the natricine snake genera *Storeria, Thamnophis,* and *Virginia;* and species of small colubrid genera such as *Sonora* and *Tantilla.*

Moreover, some modern snake genera are very difficult to separate on the basis of the osteological material that is available in the fossil record, mainly isolated vertebrae. *Coluber constrictor* and *Masticophis flagellum,* for instance, are snakes that are very different in external appearance, but which may be next to impossible to distinguish from one another on the basis of isolated vertebrae.

Another frustrating problem for the paleoherpetologist is that often when complete skeletons of Pleistocene amphibians and reptiles such as frogs or snakes are available, they are so embedded in matrix that they present only a two-dimensional view, and they usually have obscure bony processes and muscle scars. Often, a single frog ilium, unbroken and free from matrix, will be a much more diagnostic element than an entire frog embedded in matrix.

Another problem, difficult to discuss objectively, is most certainly present in Pleistocene herpetological studies. This is the ingrained idea among many paleoherpetologists that all Pleistocene herpetological material, other than certain large tortoise fossils, must belong to modern species. This bias appears to have grown out of the original negative reaction by paleoherpetologists in the late 1950s to the overzealous naming of Pleistocene species by some earlier workers such as O. P. Hay (e.g., Hay, 1908).

One who names an extinct Pleistocene species immediately becomes the object of intense scrutiny; but one who identifies a modern species on the basis of fragmentary material is seldom taken to task. It appears that this bias can be rectified only by one's keeping an open mind and by intense studies of osteological variations in modern and Pleistocene herpetological species.

Nevetheless, the limitations here are obvious. Seldom does one get an adequate sample of fossil herpetological species. Moreover, considering the intense conservation ethics that are embraced by most of us today, procuring a comparative skeletal collection that is adequate for variational studies often becomes a legal or ethical problem.

Finally, as we have seen, late Pleistocene sites are very numerous on the North American continent, whereas middle and early sites are much less common and usually are less diverse. Therefore, it could easily be argued that some earlier Pleistocene herpetological extinction events could have been missed.

But in spite of all the foregoing considerations, the Pleistocene herpetofauna of continental North America has been strikingly more stable than the avifauna or the mammalian fauna, and this has been well-documented in many parts of the continent. The following discussion will address the "why" of this Pleistocene herpetological stability.

A Look at the "Why" of Pleistocene Herpetological Stability in North America

The question of just how adaptable amphibian and reptile species are to climatic and vegetation changes is a compelling one. Students of Pleistocene paleoecology often assume that herpetological species are more sensitive to climatic and vegetational changes than are endothermic species. In fact, in the absence of other paleobotanical, palynological, and faunal information at sites, specific climatic and vegetational associations have been suggested based on the ecological requirements of modern herpetological species.

But at least in sites at relatively high elevations in the southwest (e.g., Van Devender and Mead, 1978), it has been shown that reptiles typical of the modern Sonoran Desert today existed in pinyon and pinyon-juniper vegetational associations during the late Wisconsinan. The fact that these desert reptiles could exist in vegetational associations typical of higher elevations today has been attributed to mild winter climates in the area during the Pleistocene; but these occurrences also suggest more adaptability to vegetational change in desert reptiles than was formerly believed. In fact, as has previously been discussed, the widely suggested hypothesis that desert reptiles retreated to refugia in Mexico during Pleistocene glacial episodes has been largely negated by several studies of southwestern Pleistocene herpetofaunas cited in Chapter 6.

The possibility that amphibian and reptile populations have a higher ability to adapt to vegetational and climatic changes than endothermic vertebrates is an intriguing one that needs to be substantiated by studies on adaptability in modern species as well as by continued studies of Pleistocene herpetofaunas and associated vegetational complexes. It would seem that certain herpetological species would show this adaptability more than others, as has been suggested for mammal populations in the Pleistocene (e.g., Graham and Lundelius, 1984).

Ectothermic amphibians and reptiles that have low metabolic rates, and that would be able to aestivate or hibernate during times of climatic fluctuations, would appear to have many advantages over endothermic members of the community. Continued experimental studies of aestivation and hibernation (brumation) rhythms in amphibians and reptiles should be important in establishing parameters for estimating ranges of Pleistocene adjustments among different species of amphibians and reptiles.

The small size of many North American amphibians and reptiles would appear to be an advantage to them in the face of Pleistocene habitat shrinkage or in mosaic Pleistocene communities, at least compared with the larger endotherms, especially the large mammalian herbivores. There is an analog for this, in that comparatively few very small mammalian species became extinct during the Pleistocene. It would be very interesting to have more data on how body size relates to threatened and endangered species today, and how this relates to habitat parameters.

A fascinating possibility is that few North American amphibians and reptiles were directly dependent upon the large endothermic species that became extinct during the Pleistocene, as predators, scavengers, or commensals. At the Rancholabrean Rancho La Brea site in Los Angeles, California, the scavenging or commensal birds and dung beetles that were dependent upon the large herbivorous mammals became extinct along with these mammals (Harris and Jefferson, 1985). But all the amphibian and reptile species survived into modern times with the possibility of a single tortoise bone that might represent an extinct species of *Geochelone*.

On the other hand, in Australia, the huge monitor lizard *Megalania* and the impressive constricting snake *Monambi* (Murray, 1984) were top predators, and they became extinct during the Pleistocene, presumably because many large marsupial herbivores died out. Strangely, few suggestions have been made in North American Pleistocene studies about the ecological relationships of the amphibians and reptiles to the large extinct mammals. How small amphibians and reptiles tie into the food web in African areas that support large herbivorous mammals would be interesting to know in the light of differential Pleistocene extinction rates in North America.

Parenting in birds and mammals is a part of the reproductive stress syndrome in these animals, and it would obviously be a special drain on the energy budget during times of Pleistocene climatic changes. Of course, parenting in the manner of birds and mammals is completely lacking in the noncrocodilian amphibians and reptiles of North America.

On the other hand, the high reproductive potential of many smaller amphibians and reptiles compared with the large endotherms, especially the megaherbivores (Owen-Smith, 1987), would definitely appear to favor the herpetological species in time of stress during climatic change.

For more than two decades, the hypothesis of human overkill has been advanced as an extremely important factor in the Pleistocene extinction of the large herbivorous animals and their scavengers and predators (Martin and

Wright, 1967; Martin and Klein, 1984). Holman (1959a) compared extinction percentages of amphibians and reptiles to mammals in the Williston IIIA fauna in Florida and came to the conclusion that the rise of humans might have been an important factor in the differential between extinction in the herpetological and mammalian groups. This would reflect the less desirable aspects of most amphibians and reptiles as a food resource compared with the large mammalian herbivores.

Unquestionably, the large extinct land tortoises of the genera *Geochelone* and *Gopherus* would have been an easily procured source of food for humans (e.g., at the Little Salt Spring site in Florida; Clausen et al., 1979). Moreover, the large forms of the land turtle (*Terrapene carolina*) that became extinct at the end of the Pleistocene, some of which reached a shell length of over 275 mm, would also have been easy to procure as food by humans. This might have been an important factor in the extinction of these turtle species.

In summary, the herpetofauna of North America was strikingly stable during the Pleistocene compared to the avian and mammalian fauna. There is no good evidence that any familial or generic extinctions occurred among amphibians and reptiles on the North American continent during the Pleistocene; and other than well-documented extinction on a few species of land tortoises, very little specific extinction occurred.

A provocative question then is "What attributes have allowed the herpetofauna to survive the stresses that caused such dramatic extinctions in the endothermic fauna during the Pleistocene?" Possible explanations might grow out of new and additional comparative studies involving high ability of herpetological species to adapt to climatic and vegetational changes, advantages of lower metabolic rates, hibernation and aestivation potential, size in relationship to extinction, food web relationships in megaherbivore-dominated communities, relative reproductive potentials, and less desirability as a human food resource in most species.

EPILOGUE

For some reason paleoherpetologists have not been key players in the game of interpreting of Pleistocene faunas (see Martin and Klein, 1984). Perhaps this is because there are many more mammalogists than herpetologists interested in the Pleistocene. Or it might be that, early on in paleoherpetological studies, word got out that "nothing much happened to the herpetofauna in the Pleistocene but the extinction of a few tortoises." This is unfortunate, for as we have seen, patterns of herpetological extinction, range adjustments, and adaptation to climatic change are an essential part of the interpretation of Pleistocene events.

Some things stand out above others. The North American herpetofauna became essentially modern at the family level early in the Tertiary, modern at the generic level by the end of the Miocene, and modern at the specific level by the end of the Pliocene; moreover, except for tortoises, this herpetofauna came through the epoch practically unscathed taxonomically. On the other hand, mammalian and, to a lesser extent, avian species suffered familial and generic extinction throughout the Tertiary, with a particularly devastating extinction event at the end of the Pleistocene.

Herpetological species, like mammal species, adjusted to Pleistocene events in their own individual ways rather than as parts of blocks of faunal communities marching north and south (or east and west sometimes) during Pleistocene oscillations. In other words, we find that many Pleistocene herpetofaunas, like mammalian faunas, contain extralimital species north, south, east or west of their modern ranges.

But, importantly, we also find that most herpetological species tended to stay put in the Pleistocene. This occurred all over North America but is particularly striking in the northern Appalachians. Moreover, the long-held idea that herpetological desert species migrated south to Mexican refugia in the Pleistocene has been conclusively negated by the study of Pleistocene herpetofaunas, especially in the Great Basin states and in California, thus the **practice of trying to figure out herpetological speciation events by studying modern range maps is highly questionable**.

An especially intriguing discovery is that those "desert" species that "stayed put" were able to adapt to changes from desert vegetation to pinyon-juniper forest as indicated by pollen studies. This has important implications. Many of us reasoned that herpetological species, being ectothermic, would be much more sensitive to environmental changes than birds and mammals, and thus we often tried to estimate paleoclimates on the basis of herpetological species. This concept certainly requires rethinking and refinement.

Other studies have shown that, perhaps surprisingly, herpetological species have been able to reinvade glacially obliterated habitat in a relatively short time. In Michigan, which was completely covered by a massive ice sheet that did not begin to retreat in earnest until about 14,000 years ago, an essentially modern herpetofauna had invaded southern Michigan by about 6,000 years ago.

Finally, studies of Pleistocene box turtles of the *Terrapene carolina* species complex have shown us much about how the speciation process really works. To reiterate, several subspecies of *Terrapene carolina* were alternately separated and then hybridized in various ways in the Rancholabrean Land Mammal Age, and one of these subspecies was a very large (almost "giant") box turtle. These subspecies became isolated by the end of the Pleistocene and the "giant" subspecies disappeared, leaving part of its gene pool in the population of fairly large box turtles that occurs along the Gulf Coast today.

As an epilogue, it will be interesting to see if the North American herpetofauna remains in stasis during the remainder of the Holocene. At present this seems doubtful. Yet a promising note lies in the fact that very few herpetological species have actually become extinct in modern times, with certain stressed species continuing to exist as small, isolated populations here and there. Perhaps this is one of the ways they survived then and this is the way they might be able to endure in the future.

REFERENCES

Anderson, E. (1984). Who's who in the Pleistocene: a mammalian bestiary, In P. S. Martin and R. G. Klein, eds. *Quaternary Extinctions: A Prehistoric Revolution*. Tucson: University of Arizona Press, pp. 40–89.

Auffenberg, W. (1955). Glass lizards (*Ophisaurus*) in the Pleistocene and Pliocene of Florida. *Herpetologica* **11**:133–36.

Auffenberg, W. (1956). Additional records of Pleistocene lizards from Florida. *Quarterly Journal of the Florida Academy of Sciences* **19**:157–67.

Auffenberg, W. (1958). Fossil turtles of the genus *Terrapene* in Florida. *Bulletin of the Florida State Museum, Biological Sciences* **3**:53–92.

Auffenberg, W. (1962a). A redescription of *Testudo hexagonata* Cope. *Herpetologica* **18**:25–34.

Auffenberg, W. (1962b). A new species of *Geochelone* from the Pleistocene of Texas. *Copeia,* no. 3 pp. 627–36.

Auffenberg, W. (1963a). Fossil testudine turtles of Florida genera *Geochelone* and *Floridemys*. *Bulletin of the Florida State Museum, Biological Sciences* **7**:53–97.

Auffenberg, W. (1963b). The fossil snakes of Florida. *Tulane Studies in Zoology* **10**:131–216.

Auffenberg, W. (1966). On the courtship of *Gopherus polyphemus*. Herpetologica **22**:113–17.

Auffenberg, W. (1967). Further notes on fossil box turtles from Florida. *Copeia,* no. 2 pp. 319–25.

Auffenberg, W. (1974). Checklist of fossil land tortoises (Testudinidae). *Bulletin of the Florida State Museum, Biological Sciences* **18**:121–251.

Auffenberg, W. (1988). A new species of *Geochelone* (Testudinata:Testudinidae) from the Pleistocene of Florida (U.S.A.). *Acta Zoologica Cracoviensia* **31**:591–604.

Auffenberg, W., and Milstead, W. W. (1965). Reptiles in the Quaternary of North America. In H.E. Wright and D. G. Frey, eds. *The Quaternary of the United States*. Princeton: Princeton University Press, pp. 557–68.

Axelrod, D. I. (1967). Quaternary extinctions of large mammals. *University of California Publications of the Geological Society* **74**:1–42.

Banks, R. C., McDiarmid, R. W., and Gardner, A. L., eds. (1987). Checklist of vertebrates of the United States, the U. S. territories, and Canada. *United States Department of the Interior Fish and Wildlife Service Resource Publication* **166**:1–79.

Bell, C. J., and Dundas, R. G. (1993). Fossil lizards from Rancho La Brea in the Collections of the University of California Museum of Paleontology. *Paleobios* **15**:14–17.

Bentley, C. C., and Knight, J. L. (1994). The oldest spotted turtle: *Clemmys guttata* (Testudines:Emydidae) from the late Pleistocene (Rancholabrean) Ardis Local Fauna, Dorchester County, South Carolina. *South Carolina Geology* (in press).

Brattstrom, B. H. (1953a). Records of Pleistocene reptiles from California. *Copeia,* no. 3 pp. 174–79.

Brattstrom, B. H. (1953b). The amphibians and reptiles from Rancho La Brea. *Transactions of the San Diego Society of Natural History* **11**:365–92.

Brattstrom, B. H. (1953c). Records of Pleistocene reptiles and amphibians from Florida. *Quarterly Journal of the Florida Academy of Sciences* **16**:243–48.

Brattstrom, B. H. (1954a). Amphibians and reptiles from Gypsum Cave, Nevada. *Bulletin of the Southern California Academy of Sciences* **53**:8–12.

Brattstrom, B. H. (1954b). The fossil pit-vipers (Reptilia:Crotalidae) of North America. *Transactions of the San Diego Society of Natural History* **12**:31–46.

Brattstrom, B. H. (1955a). Pliocene and Pleistocene amphibians and reptiles from southeastern Arizona. *Journal of Paleontology* **29**:150–54.

Brattstrom, B. H. (1955b). Small herpetofauna from the Pleistocene of Carpenteria, California. *Copeia,* no. 2 pp. 138–39.

Brattstrom, B. H. (1958a). New records of Cenozoic amphibians and reptiles from California. *Bulletin of the Southern California Academy of Sciences* **57**:5–12.

Brattstrom, B. H. (1958b). Additions to the Pleistocene herpetofauna of Nevada. *Herpetologica* **14**:36.

Brattstrom, B. H. (1961). Some new fossil tortoises from western North America with remarks on zoogeography and paleoecology of tortoises. *Journal of Paleontology* **35**:543–60.

Brattstrom, B. H. (1964). Amphibians and reptiles from cave deposits in south-central New Mexico. *Bulletin of the Southern California Academy of Sciences* **63**:93–103.

Brattstrom, B. H. (1967). A succession of Pliocene and Pleistocene snake faunas from the High Plains of the United States. *Copeia,* no. 1 pp. 188–202.

Brattstrom, B. H. (1976). A Pleistocene herpetofauna from Smith Creek Cave, Nevada. *Bulletin of the Southern California Academy of Sciences* **75**:283–84.

Brown, B. (1908). The Conard Fissure, a Pleistocene bone deposit in northern Arkansas: with descriptions of two new genera and twenty new species of animals. *Memoirs of the American Museum of Natural History* **9**:155–208.

Campbell, J. A., and Vannini, J. P. (1988). A new subspecies of beaded lizard, *Heloderma horridum,* from the Motagua Valley of Guatemala. *Journal of Herpetology* **22**:457–68.

Chantell, C. J. (1966). Late Cenozoic hylids from the Great Plains. *Herpetologica* **22**:259–64.

Churcher, C. S., and Dods, R. R. (1979). *Ochotona* and other vertebrates of possible Illinoian age from Kelso Cave, Halton County, Ontario. *Canadian Journal of Earth Science* **16**:1613–20.

Churcher, C. S., and Peterson, R. L. (1982). Chronologic and environmental implications of a new genus of fossil deer from late Wisconsinan deposits at Toronto, Canada. *Quaternary Research* **18**:184–93.

Clausen, C. J., Cohen, A. D., Emiliani, C., Holman, J. A., and Stipp, J. J. (1979). Little Salt Spring, Florida: a unique underwater site. *Science* **203**:609–14.

Conant, R., and Collins, J. T. (1991). *A Field Guide to Reptiles and Amphibians: Eastern and Central North America*. Boston: Houghton Mifflin.

Cooke, C. M. (1945). Geology of Florida. *Florida Geological Survey Bulletin* **29**:1–339.

Daeschler, E., Spamer, E. E., and Parris, D. C. (1993). Review and new data on the Port Kennedy local fauna (late Irvingtonian), Valley Forge National Historical Park, Montgomery County, Pennsylvania. *The Mosasaur* **5**:23–41.

Daly, E. (1992). A list, bibliography and index of the fossil vertebrates of Mississippi. *Mississippi Office of Geology Bulletin* **128**:1–47.

Davis, L. C. (1973). The herpetofauna of Peccary Cave, Arkansas. Masters thesis, University of Arkansas, Fayetteville.

DeFauw, S. L., and Shoshani, J. (1991). *Rana catesbeiana* and *R. clamitans* from the late Pleistocene of Michigan. *Journal of Herpetology* **25**:95–99.

Delcourt, H. R., Delcourt, P. A., and Webb, T., III. (1983). Dynamic plant ecology: the spectrum of vegetational change in space and time. *Quaternary Science Reviews* **1**:153–75.

Dobie, J. L., and Jackson. D. R. (1979). First fossil record for the diamondback terrapin, *Malaclemys terrapin* (Emydidae), and comments on the fossil record of *Chrysemys nelsoni* (Emydidae). *Herpetologica* **35**:139–45.

Dorr, J. A., and Eschman, D. F. (1970). *Geology of Michigan*. Ann Arbor: University of Michigan Press.

Dowling, H. G. (1958). Pleistocene snakes of the Ozark Plateau. *American Museum Novitates* **1882**:1–9.

Dowling, H. G. (1959). Classification of Serpentes: a critical review. *Copeia,* no. 1 pp. 38–52.

Dunbar, C. O. (1949). *Historical Geology*. New York: John Wiley and Sons.

Estes, R. (1970). Origin of the Recent North American lower vertebrate fauna: an inquiry into the fossil record. *Forma et Functio* **3**:139–63.

Estes. R. (1981). Gymnophiona, Caudata. Part 2. *Handbuch der Paläoherpetologie*. Stuttgart: Fischer Verlag.

Estes, R. (1983). Sauria terrestria, Amphisbaenia. Part 10A. *Handbuch der Paläoherpetologie*. Stuttgart: Fischer Verlag.

Estes, R., and Baez, A. (1985). Herpetofauna of North and South America during the late Cretaceous and Cenozoic: evidence for interchange?, In F. G. Stehli and S. D. Webb, eds. *The Great American Biotic Interchange*. New York: Plenum, pp. 139–96.

Etheridge, R. (1958). Pleistocene lizards of the Cragin Quarry fauna of Meade County, Kansas. *Copeia,* no. 2 pp. 94–101.

Etheridge, R. (1960a). The slender glass lizard, *Ophisaurus attenuatus,* from the Pleistocene (Illinoian glacial) of Oklahoma. *Copeia,* no. 1 pp. 46–47.

Etheridge, R. (1960b). Additional notes on lizards of the Cragin Quarry fauna. *Papers of the Michigan Academy of Science, Arts and Letters* **45**:113–17.

Etheridge, R. (1961). Late Cenozoic glass lizards (*Ophisaurus*) from the southern Great Plains. *Herpetologica* **17**:179–86.

Fay, L. P. (1984). Mid-Wisconsinan and mid-Holocene herpetofaunas of eastern North America: a study in minimal contrast. *Carnegie Museum of Natural History Special Publication* **8**:14–19.

Fay, L. P. (1986). Wisconsinan herpetofaunas of the central Appalachians. *Virginia Division of Mineral Resources Publication* **75**:126–28.

Fay, L. P. (1988). Late Wisconsinan Appalachian herpetofaunas: relative stability in the midst of change. *Annals of Carnegie Museum* **57**:189–220.

Fay, L. P. (1991). First record of tortoises (Reptilia:Testudinidae) from Rancho La Brea. *Abstract 8 of the Annual Meeting of the Southern California Academy of Sciences,* Los Angeles, May 10–11.

Fay, L. P. (n.d.) Guy Wilson cave, Sullivan County, Tennessee, Woodfordian (full-glacial) 19,700 plus or minus 600 (I-4163). Letter to J. A. Holman, May 12, 1993.

Fay, L. P. (n.d.) Hanover Quarry No. 1 fissure, Adams County, Pennsylvania, Early (?) Irvingtonian Land Mammal "Age." Letter to J. A. Holman, May 12, 1993.

Flint, R. F. (1971). *Glacial and Quaternary Geology*. New York: John Wiley and Sons.

Foley, R. L. (1984). Late Pleistocene (Woodfordian) vertebrates from the driftless area of southwestern Wisconsin, the Moscow Fissure local fauna. *Illinois State Museum Reports of Investigations* **39**:1–50.

Forbes, E. (1846). On the connection between the existing fauna and flora of the British Isles, and the geological changes which have affected their area, especially during the epoch of the Northern Drift. *Memoirs of the Great Britain Geological Survey* **1**:336–432.

Ford, K. M., III (1992). Herpetofauna of the Albert Ahrens local fauna (Pleistocene:Irvingtonian), Nebraska. Masters thesis, Michigan State University, East Lansing.

Galbreath, E. C. (1938). Post-glacial fossil vertebrates from east-central Illinois. *Field Museum of Natural History Geological Series* **4**:303–13.

Galbreath, E. C. (1948). Pliocene and Pleistocene records of fossil turtles from western Kansas and Oklahoma. *University of Kansas Publications, Museum of Natural History* **1**:281–84.

Gehlbach, F. R. (1965). Amphibians and reptiles from the Pliocene and Pleistocene of North America: a chronological summary and selected bibliography. *Texas Journal of Science* **17**:56–70.

Gilmore, C. W. (1938). Fossil snakes of North America. *Geological Society of America Special Paper* **9**:1–96.

Goin, C. J., and Auffenberg, W. (1955). The fossil salamanders of the family Sirenidae. *Bulletin of the Museum of Comparative Zoology* **113**:497–514.

Graham, R. W. (1976). Late Wisconsinan mammalian faunas and environmental gradients of the eastern United States. *Paleobiology* **2**:343–50.

Graham, R. W. (1979). Paleoclimates and late Pleistocene faunal provinces in North America. In R. L. Humphrey and D. Stanford, eds. *Pre-Llano Cultures of the Americas: Possibilities and Paradoxes*. Washington D.C.: Washington Anthropological Society, pp. 49–69.

Graham, R. W., Holman, J. A., and Parmalee, P. W. (1983). Taphonomy and paleoecology of the Christensen Bog mastodon bone bed, Hancock County, Indiana. *Illinois State Museum Reports of Investigations* **38**:1–29.

Graham, R. W., and Lundelius, E. L., Jr. (1984). Coevolutionary disequilibrium and Pleistocene extinctions. In P. S. Martin and R. G. Klein, eds. *Quaternary Extinctions: A Prehistoric Revolution*. Tucson: University of Arizona Press, pp. 223–49.

Grayson, D. S. (1984). Nineteenth century explanations of Pleistocene extinctions: a review and analysis. In P. S. Martin and R. G. Klein, eds. *Quaternary Extinctions: A Prehistoric Revolution*. Tucson: University of Arizona Press, pp. 5–39.

Guilday, J. E. (1962). The Pleistocene local fauna of the Natural Chimneys, Augusta County, Virginia. *Annals of Carnegie Museum* **36**:87–122.

Guilday, J. E. (1977). Sabertooth cat, *Smilodon floridanus* (Leidy), and associated fauna from a Tennessee cave (40 Dv 40), the First American Bank Site. *Journal of the Tennessee Academy of Science* **52**:84–94.

Guilday, J. E., Hamilton, H. W., and McCrady, A. D. (1966). The bone breccia of Bootlegger Sink, York County, PA. *Annals of Carnegie Museum* **38**:145–63.

Guilday, J. E., Hamilton, H. W., and McCrady, A. D. (1969). The Pleistocene vertebrate fauna of Robinson Cave, Overton County, Tennessee. *Palaeovertebrata Montpellier* **2**:25–75.

Guilday, J. E., Martin, P. S., and McCrady, A. D. (1964). New Paris No. 4: a late Pleistocene cave deposit in Bedford County, Pennsylvania. *Bulletin of the National Speleological Society* **26**:122–94.

Gut, H. J., and Ray, C. E. (1963). The Pleistocene vertebrate fauna of Reddick, Florida. *Quarterly Journal of the Florida Academy of Sciences* **26**:315–28.

Hallock, L. A., Holman, J. A., and Warren, M. R. (1990). Herpetofauna of the Ipswichian Interglacial bed of the Itteringham Gravel Pit, Norfolk, England. *Journal of Herpetology* **24**:33–39.

Hansen, M. C. (1992, Spring). Indian Trail Caverns: a window on Ohio's Pleistocene bestiary. *Ohio Geology,* pp. 1, 3.

Harris, A. H. (1987). Reconstruction of mid-Wisconsinan environments in southern New Mexico. *National Geographic Research* **3**:142–51.

Harris, J. M., and Jefferson, G. T. (1985). *Rancho La Brea: Treasures of the Tar Pits.* Los Angeles: Natural History Museum Publication.

Hawksley, O. (1986). Remains of Quaternary vertebrates from Ozark caves and other miscellaneous sites. *Missouri Speleology* **26**:1–67.

Hay, O. P. (1908). The fossil turtles of North America. *Carnegie Institute of Washington Publication* **75**:1–568.

Hay, O. P. (1917). On a collection of vertebrates made by Dr. F. W. Cragin in the *Equus* beds of Kansas. *Kansas University Science Bulletin* **10**:39–51.

Hecht, M. K. (1956). A new xantusiid lizard from the Eocene of Wyoming. *American Museum of Natural History Novitates* **1174**:1–8.

Hibbard, C. W. (1960). An interpretation of Pliocene and Pleistocene climates in North America. *President's Address, 62nd Annual Report of the Michigan Academy of Science, Arts and Letters*, pp. 5–30.

Hill, W. H. (1971). Pleistocene snakes from a cave in Kendall County, Texas. *Texas Journal of Science* **21**:209–16.

Holman, J. A. (1958). The Pleistocene herpetofauna of Saber-tooth Cave, Citrus County, Florida. *Copeia,* no. 4 pp. 276–80.

Holman, J. A. (1959a). Amphibians and reptiles from the Pleistocene (Illinoian) of Williston, Florida. *Copeia,* no. 2 pp. 96–102.

Holman, J. A. (1959b). A Pleistocene herpetofauna near Orange Lake, Florida. *Herpetologica* **15**:121–25.

Holman, J. A. (1962a). Additional records of Florida Pleistocene amphibians and reptiles. *Herpetologica* **18**:115–19.

Holman, J. A. (1962b). A Texas Pleistocene herpetofauna. *Copeia,* no. 2 pp. 255–61.

Holman, J. A. (1963). Late Pleistocene amphibians and reptiles of the Clear Creek and Ben Franklin local faunas of Texas. *Journal of the Graduate Research Center, Southern Methodist University* **31**:152–67.

Holman, J. A. (1964). Pleistocene amphibians and reptiles from Texas. *Herpetologica* **20**:73–83.

Holman, J. A. (1965a). Pleistocene snakes from the Seymour Formation of Texas.

Copeia, no. 1 pp. 102–4.

Holman, J. A. (1965b). A late Pleistocene herpetofauna from Missouri. *Transactions of the Illinois Academy of Science* **58**:190–94.

Holman, J. A. (1965c). A huge Pleistocene box turtle from Texas. *Quarterly Journal of the Florida Academy of Sciences* **28**:245–48.

Holman, J. A. (1965d). A small Pleistocene herpetofauna from Houston, Texas. *Texas Journal of Science* **27**:418–23.

Holman, J. A. (1966a). Some Pleistocene turtles from Illinois. *Transactions of the Illinois State Academy of Sciences* **59**:214–16.

Holman, J. A. (1966b). The Pleistocene herpetofauna of Miller's Cave, Texas. *Texas Journal of Science* **28**:272–77.

Holman, J. A. (1967). A Pleistocene herpetofauna from Ladds, Georgia. *Bulletin of the Georgia Academy of Science* **25**:154–66.

Holman, J. A. (1968). A Pleistocene herpetofauna from Kendall County, Texas. *Quarterly Journal of the Florida Academy of Sciences* **31**:165–72.

Holman, J. A. (1969a). Pleistocene amphibians from a cave in Edwards County, Texas. *Texas Journal of Science* **31**:63–67.

Holman, J. A. (1969b). The Pleistocene amphibians and reptiles of Texas. *Publications of the Museum, Michigan State University, Biological Series* **4**:163–92.

Holman, J. A. (1969c). Herpetofauna of the Pleistocene Slaton local fauna of Texas. *Southwestern Naturalist* **14**:203–12.

Holman, J. A. (1970). A Pleistocene herpetofauna from Eddy County, New Mexico. *Texas Journal of Science* **21**:29–39.

Holman, J. A. (1971). Herpetofauna of the Sandahl local fauna (Pleistocene:Illinoian) of Kansas. *Contributions from the Museum of Paleontology, the University of Michigan* **23**:349–55.

Holman, J. A. (1972). Herpetofauna of the Kanapolis local fauna (Pleistocene:Yarmouth) of Kansas. *Michigan Academician* **5**:87–98.

Holman, J. A. (1974). A late Pleistocene herpetofauna from southwestern Missouri. *Journal of Herpetology* **8**:343–46.

Holman, J. A. (1975). Michigan's fossil vertebrates. *Publications of the Museum, Michigan State University, Educational Series* **2**:1–54.

Holman, J. A. (1976). Paleoclimatic implications of "ecologically incompatible" herpetological species (late Pleistocene:southeastern United States). *Herpetologica* **32**:290–95.

Holman, J. A. (1977a). The Pleistocene (Kansan) herpetofauna of Cumberland Cave, Maryland. *Annals of Carnegie Museum* **46**:157–72.

Holman, J. A. (1977b). America's northernmost Pleistocene herpetofauna (Java, north-central South Dakota). *Copeia,* no. 1 pp. 191–93.

Holman, J. A. (1978). The late Pleistocene herpetofauna of Devil's Den sinkhole, Levy County, Florida. *Herpetologica* **34**:228–37.

Holman, J.A. (1979). Herpetofauna of the Nash local fauna (Pleistocene:Aftonian) of Kansas. *Copeia,* no. 4 pp. 747–49.

Holman, J. A. (1980). Paleoclimatic implications of Pleistocene herpetofaunas of eastern and central North America. *Transactions of the Nebraska Academy of Sciences* **8**:131–40.

Holman, J. A (1981). A review of North American Pleistocene snakes. *Publications of the Museum, Michigan State University, Paleontological Series* **1**:261–306.

Holman, J. A. (1982). The Pleistocene (Kansan) herpetofauna of Trout Cave, West

Virginia. *Annals of Carnegie Museum* **51**:391–404.

Holman, J. A. (1984). Herpetofaunas of the Duck Creek and Williams local faunas (Pleistocene:Illinoian) of Kansas. *Special Publication of the Carnegie Museum of Natural History* **8**:20–38.

Holman, J. A. (1985a). Herpetofauna of Ladds Quarry. *National Geographic Research* **1**:423–36.

Holman, J. A. (1985b). New evidence on the status of Ladds Quarry. *National Geographic Research* **1**:569–70.

Holman, J. A. (1986a). The known herpetofauna of Virginia poses a dilemma. *Virginia Division of Mineral Resources Publication* **75**:36–42.

Holman, J. A. (1986b). Turtles from the late Wisconsinan of west-central Ohio. *American Midland Naturalist* **116**:213–14.

Holman, J. A. (1986c). Butler Spring herpetofauna of Kansas (Pleistocene:Illinoian) and its climatic significance. *Journal of Herpetology* **20**:568–70.

Holman, J. A. (1987a). Climatic significance of a late Illinoian herpetofauna from southwestern Kansas. *Contributions from the Museum of Paleontology, the University of Michigan* **27**:129–41.

Holman, J. A. (1987b). Snakes from the Robert local fauna (late Wisconsinan) of Meade County, Kansas. *Contributions from the Museum of Paleontology, the University of Michigan* **27**:143–50.

Holman, J. A. (1988). The status of Michigan's Pleistocene herpetofauna. *Michigan Academician* **20**:125–32.

Holman, J. A. (1989). Additional herpetological records from the middle Pleistocene (Cromerian interglacial) freshwater bed, West Runton, Norfolk. *British Herpetological Society Bulletin* **27**:9–12.

Holman, J. A. (1991a). New records of Michigan Pleistocene vertebrates with comments on the Mason-Quimby Line. *Michigan Academician* **23**:273–83.

Holman, J. A. (1991b). North American Pleistocene herpetofaunal stability and its impact on the interpretation of modern herpetofaunas: an overview. *Illinois State Museum Scientific Papers* **23**:227–35.

Holman, J. A. (1992a). Late Quaternary herpetofauna of the central Great Lakes region, U.S.A.: zoogeographical and paleoecological implications. *Quaternary Science Reviews* **11**:345–51.

Holman, J. A. (1992b). Patterns of herpetological re-occupation of post-glacial Michigan: amphibians and reptiles come home. *Michigan Academician* **24**:453–66.

Holman, J. A. (1994). Herpetofauna of the Sheriden Pit late Pleistocene site, Wyandot County, Ohio. In preparation.

Holman, J. A., Bell, G., and Lamb, J. (1990). A late Pleistocene herpetofauna from Bell Cave, Alabama. *Herpetological Journal* **1**:521–29.

Holman, J. A., and Clausen, C. J. (1984). Fossil vertebrates associated with paleo-Indian artifacts at Little Salt Spring, Florida. *Journal of Vertebrate Paleontology* **4**:146–54.

Holman, J. A., and Fisher, D. C. (1993). Late Pleistocene turtle remains (Reptiles: Testudines) from southern Michigan. *Michigan Academician* **25**:491–99.

Holman, J. A., Fisher, D. C., and Kapp, R. O. (1986). Recent discoveries of fossil vertebrates in the Lower Peninsula of Michigan. *Michigan Academician* **18**:431–63.

Holman, J. A., and Grady, F. (1987). Herpetofauna of New Trout Cave. *National Geographic Research* **3**:305–17.

Holman, J. A., and Grady, F. (1989). The fossil herpetofauna (Pleistocene:Irvingtonian) of Hamilton Cave, Pendleton County, West Virginia. *National Speleological Society Bulletin* **51**:34–41.

Holman, J. A., and Harding, J. H. 1977. Michigan's turtles. *Publications of the Museum, Michigan State University Educational Bulletin* **3**:1–40.

Holman, J. A., and McDonald, J. N. (1986). A late Quaternary herpetofauna from Saltville, Virginia. *Brimleyana* **12**:85–100.

Holman, J. A., and Richards, R. L. (1981). Late Pleistocene occurrence in southern Indiana of the smooth green snake, *Opheodrys vernalis*. *Journal of Herpetology* **15**:123–25.

Holman, J. A., and Richards, R. L. (1993). Herpetofauna of the Prairie Creek site, Daviess County, Indiana. *Proceedings of the Indiana Academy of Science* **102**:115–31.

Holman, J. A., and Schloeder, M. E. (1991). Fossil herpetofauna of the Lisco C quarries (Pliocene:early Blancan) of Nebraska. *Transactions of the Nebraska Academy of Sciences* **18**:19–29.

Holman, J. A., and Voorhies, M. R. (1985). *Siren* (Caudata:Sirenidae) from the Barstovian Miocene of Nebraska. *Copeia,* no. 1 pp. 264–66.

Holman, J. A., and Winkler, A. J. (1987). A mid-Pleistocene (Irvingtonian) herpetofauna from a cave in southcentral Texas. *Pearce-Sellards Series, Texas Memorial Museum, the University of Texas at Austin* **44**:1–17.

Hood, C. H., and Hawksley, O. (1975). A Pleistocene fauna from Zoo Cave, Taney County, Missouri. *Missouri Speleology* **15**:1–42.

Hudson, D. M., and Brattstrom, B. H. (1977). A small herpetofauna from the late Pleistocene of Newport Beach mesa, Orange County, California. *Bulletin of the Southern California Academy of Sciences* **76**:16–20.

Hulbert, R. C., and Morgan, G. S. (1989). Stratigraphy, paleoecology, and vertebrate fauna of the Leisey Shell Pit local fauna, early Pleistocene (Irvingtonian) of southwestern Florida. *Papers in Florida Paleontology* **2**:1–19.

Imbrie, J. (1977). On the underlying causes of Quaternary climatic change. *Abstracts of the 10th INQUA Congress, August 1977.* Birmingham, U.K., p. 221.

Jackson, C. G., and Kaye, J. M., Jr. (1974a). Occurrence of box turtles *Terrapene* (Testudines:Testudinidae) in the Pleistocene of Mississippi. *Herpetologica* **30**:11–13.

Jackson, C. G., and Kaye, J. M., Jr. (1974b). The occurrence of Blanding's turtle, *Emydoidea blandingii*, in the late Pleistocene of Mississippi. *Herpetologica* **30**:417–19.

Jackson, C. G., and Kaye, J. M., Jr. (1975). Giant tortoises in the late Pleistocene of Mississippi. *Herpetologica* **31**:421.

Jackson, D. R. (1975). A Pleistocene *Graptemys* (Reptilia:Testudines) from the Santa Fe River of Florida. *Herpetologica* **31**:213–19.

Jackson, D. R. (1978). Evolution and fossil record of the chicken turtle *Deirochelys*, with a re-evaluation of the genus. *Tulane Studies in Zoology* **20**:35–55.

Johnson, E. (1974). Zooarcheology and the Lubbock Lake site. *Museum Bulletin of the West Texas Museum Association* **15**:107–22.

Johnson, E. (1987). Vertebrate remains. In E. Johnson ed. *Late Quaternary Studies on the Southern High Plains.* College Station: Texas A&M University Press, pp. 49–89.

Kasper, S., and Parmley, D. (1990). A late Pleistocene herpetofauna from the lower Texas Panhandle. *Texas Journal of Science* **42**:289–94.

Kaye, J. M. (1974). Pleistocene sediment and vertebrate fossil associations in the

Mississippi Black Belt. Ph.D. dissertation, Louisiana State University, Baton Rouge.

King, J. E., and Saunders, J. J. (1986). *Geochelone* in Illinois and the Illinoian-Sangamonian vegetation of the type region. *Quaternary Research* **25**:89–99.

Klein, R. G. (1984). Mammalian extinction and Stone Age people in Africa, p. 553–73. In P. S. Martin and R. G. Klein, eds. *Quaternary Extinctions: A Prehistoric Revolution*. Tucson: University of Arizona Press, pp. 553–73.

Klippel, W. E., and Parmalee, P. W. (1982). The paleontology of Cheek Bend Cave. *Phase II Report to the Tennessee Valley Authority*, pp. 1–249.

Kurtén B., and Anderson, E. (1980). *Pleistocene Mammals of North America*. New York: Columbia University Press.

LaDuke, T. C. (1991a). The fossil snakes of Pit 91, Rancho La Brea, California. *Natural History Museum of Los Angeles County, Contributions in Science* **424**:1–28.

LaDuke, T. C. (1991b). First record of salamander remains from Rancho La Brea. *Abstract 7 of the Annual Meeting of the Southern California Academy of Sciences,* Los Angeles, May 10–11.

Lundelius, E. L. (1972). Fossil vertebrates from the late Pleistocene Ingleside fauna, San Patricio County, Texas. *Bureau of Economic Geology Report of Investigations* **77**:1–74.

Lundelius, E. L., Graham, R. W., Anderson, E., Guilday, J., Holman, J. A., Steadman, D. W., and Webb, S. D. (1983). Terrestrial vertebrate faunas. In H. E. Wright, ed. *Late Quaternary Environments of the United States,* Vol 1: *The Late Pleistocene,* S. Porter, ed. Minneapolis: University of Minnesota Press, pp. 311–53.

Lyell, C. (1839). *Nouveaux elements de Geologie*. Paris: Pitois-Levrault.

Lynch, J. D. (1964). Additional hylid and leptodactylid remains from the Pleistocene of Texas and Florida. *Herpetologica* **20**:141–42.

Lynch, J. D. (1965). The Pleistocene amphibians of Pit II, Arredondo, Florida. *Copeia,* no. 1 pp. 72–77.

Lynch, J. D. (1966). Additional treefrogs (Hylidae) from the North American Pleistocene. *Annals of Carnegie Museum* **38**:265–71.

MacStalker, A. (1977). The megablocks, or giant eratics of the Canadian prairies. *Abstracts of the 10th INQUA Congress, August, 1977.* Birmingham, U.K., p. 32.

Martin, P. S. (1967). Prehistoric overkill. In P. S. Martin and H. E. Wright, eds. *Pleistocene Extinctions: The Search for a Cause*. New Haven: Yale University Press, pp. 75–120.

Martin, P. S. (1973). The discovery of America. *Science* **179**:969–74.

Martin, P. S. (1984). Prehistoric overkill: the global process. In P. S. Martin and R. G. Klein, eds. *Quaternary Extinctions: A Prehistoric Revolution*. Tucson: University of Arizona Press, pp. 354–403.

Martin, P. S., and Klein, R. G., eds. (1984). *Quaternary Extinctions: A Prehistoric Revolution*. Tucson: University of Arizona Press.

Martin, P. S., and Wright, H. E., eds. (1967). *Pleistocene Extinctions: The Search for a Cause*. New Haven: Yale University Press.

Martin, R. A. (1974). Fossil vertebrats from the Haile XIVA fauna, Alachua County. In Webb, S. D., ed. *Pleistocene Mammals of Florida*. Gainesville: University of Florida Press, pp. 100–113.

Matsch, C. L. (1976). *North America and the Great Ice Age*. New York: McGraw-Hill.

McCrady, E. (1954). A new species of *Gyrinophilus* (Plethodontidae) from Tennessee caves. *Copeia,* no. 3 pp. 200–206.

McDonald, H. G., and Anderson, E. (1975). A late Pleistocene vertebrate fauna from southeastern Idaho. *Tebiwa* **18**:20–37.

Mead, J. I. (1985). Paleontology of Hidden Cave: amphibians and reptiles. In D. H. Thomas, ed., The archaeology of Hidden Cave, Nevada. *Anthropological Papers of the American Museum of Natural History* **61**:162–70.

Mead, J. I., Heaton, T. H., and Mead, E. M. (1989). Late Quaternary reptiles from two caves in the east-central Great Basin. *Journal of Herpetology* **23**:186–89.

Mead, J. I., and Meltzer, D. J. (1984). North American late Quaternary extinctions and the radiocarbon record. In P. S. Martin and R. G. Klein, eds. *Quaternary Extinctions: A Prehistoric Revolution*. Tucson: University of Arizona Press, pp. 440–50.

Mead, J. I., Roth, E. L., Van Devender, T. R., and Steadman, D. W. (1984). The late Wisconsinan vertebrate fauna from Deadman Cave, southern Arizona. *Transactions of the San Diego Society of Natural History* **20**:247–76.

Mead, J. I., Thompson, R. S., and Van Devender, T. R. (1982). Late Wisconsinan and Holocene fauna from Smith Creek Canyon, Snake Range, Nevada. *Transactions of the San Diego Society of Natural History* **20**:1–26.

Mead, J. I., Van Devender, T. R., Cole, K. L., and Wake, D. B. (1985). Late Pleistocene vertebrates from a packrat midden in south-central Sierra Nevada, California. *Current Research in the Pleistocene* **2**:107–8.

Mecham, J. S. (1959). Some Pleistocene amphibians and reptiles from Friesenhahn Cave, Texas. *Southwestern Naturalist* **3**:17–27.

Meltzer, D. J., and Mead, J. I. (1983). The timing of late Pleistocene mammalian extinctions in North America. *Quaternary Research* **19**:130–35.

Mehl, M. G. (1962). Missouri's Ice Age animals. *State of Missouri Division of Geological Survey and Water Resources Educational Series* **1**:1–104.

Meylan, P. A. (1982). The squamate reptiles of the Inglis IA fauna (Irvingtonian: Citrus County, Florida). *Bulletin of the Florida State Museum, Biological Sciences* **27**:1–85.

Meylan, P. A. (1984). A history of fossil amphibians and reptiles in Florida. *Plaster Jacket* **44**:1–28.

Miller, L. (1942). A Pleistocene tortoise from the McKittrick Asphalt. *Transactions of the San Diego Society of Natural History* **9**:439–42.

Miller, M. D., Jr. (1992). Analysis of fossil salamanders from Cheek Bend Cave, Maury County, Tennessee. Masters thesis, Appalachian State University, Boone, North Carolina.

Miller, W. D. (1971). Pleistocene vertebrates of the Los Angeles Basin and vicinity (exclusive of Rancho La Brea). *Natural History Museum of Los Angeles County Science Bulletin* **10**:1–24.

Milstead, W. W. (1956). Fossil turtles of Friesenhahn Cave, Texas, with the description of a new species of *Testudo*. *Copeia,* no. 3 pp. 162–71.

Milstead, W. W. (1967). Fossil box turtles (*Terrapene*) from central North America and box turtles of eastern Mexico. *Copeia,* no. 1 pp. 168–79.

Moodie, K. B., and Van Devender, T. R. (1977). Additional late Pleistocene turtles from Jones Spring, Hickory County, Missouri. *Herpetologica* **33**:87–90.

Murray, P. (1984). Extinctions down under: a bestiary of extinct Australian late Pleistocene monotremes and marsupials, In P. S. Martin and R. G. Klein, eds. *Quaternary Extinctions: A Prehistoric Revolution*. Tucson: University of Arizona Press, pp. 600–628.

Naylor, B. G. (1978) The earliest known *Necturus* from the Paleocene Ravenscrag

Formation of Saskatchewan. *Journal of Herpetology* **12**:565–69.

Naylor, B. G. (1981). Cryptobranchid salamanders from the Paleocene and Miocene of Saskatchewan. *Copeia,* no. 1 pp. 76–86.

Norris, K. S., and Kavanau (1966). The burrowing of the western shovel-nosed snake, *Chionactis occipitalis* Hallowell, and the under-sand environment. *Copeia,* no. 4 pp. 650–64.

Oelrich, T. M. (1953). A new boxturtle from the Pleistocene of southwestern Kansas. *Copeia,* no. 1 pp. 33–38.

Owen-Smith, N. (1987). Pleistocene extinctions: the pivotal role of megaherbivores. *Paleobiology* **13**:351–62.

Parmalee, P. W., and Klippel, W. E. (1981). Remains of the wood turtle *Clemmys insculpta* (Le Conte) from a late Pleistocene deposit in middle Tennessee. *American Midland Naturalist* **105**:413–16.

Parmalee, P. W., and Oesch, R. D. (1972). Pleistocene and Recent faunas from the Brynjulfson Caves, Missouri. *Illinois State Museum Reports of Investigations* **25**:1–52.

Parmalee, P. W., Oesch, R. D., and Guilday, J. E. (1969). Pleistocene and Recent vertebrate faunas from Crankshaft Cave, Missouri. *Illinois State Museum Reports of Investigations* **14**:1–37.

Parmley, D. (1986). Herpetofauna of the Rancholabrean Schulze Cave local fauna of Texas. *Journal of Herpetology* **20**:1–10.

Parmley, D. (1988a). Additional Pleistocene amphibians and reptiles from the Seymour Formation, Texas. *Journal of Herpetology* **22**:82–87.

Parmley, D. (1988b). Late Pleistocene anurans from Fowlkes Cave, Culberson County, Texas. *Texas Journal of Science* **40**:357–58.

Parmley, D. (1990). Late Pleistocene snakes from Fowlkes Cave, Culberson County, Texas. *Journal of Herpetology* **24**:266–74.

Patton, T. H. (1963). Fossil vertebrates from Miller's Cave, Llano County, Texas. *Bulletin of the Texas Memorial Museum* **7**:1–41.

Pilny, J. J., and Morgan, A. V. (1987). Paleoentomology and paleoecology of a possible Sangamonian site near Innerkip, Ontario. *Quaternary Research* **28**:157–74.

Preston, R. E. (1966). Turtles of the Gilliland faunule from the Pleistocene of Knox County, Texas. *Papers of the Michigan Academy of Science, Arts and Letters* **71**:221–39.

Preston, R. E. (1971). Pleistocene turtles from the Arkalon local fauna of southwestern Kansas. *Journal of Herpetology* **5**:208–11.

Preston, R. E. (1979). Late Pleistocene cold-blooded vertebrate faunas from the mid-continental United States, I. Reptilia; Testudines, Crocodilia. *University of Michigan Museum of Paleontology, Papers on Paleontology* **19**:1–53.

Preston, R. E., and McCoy, C. J. (1971). The status of *Emys twentei* Taylor (Reptilia: Testudinidae) based on new fossil records from Kansas and Oklahoma. *Journal of Herpetology* **5**:23–30.

Rage, J. C. (1984). Serpentes. Part 11. *Handbuch der Paläoherpetologie*. Stuttgart: Fischer Verlag.

Repenning, C. A. (1987). Biochronology of the microtine rodents of the United States. In M. O. Woodburne, ed. *Cenozoic Mammals of North America*. Berkeley: University of California Press, pp. 236–68.

Richards, R. L. (1984). The Pleistocene vertebrate collection of the Indiana State Museum, with a list of the extinct and extralocal Pleistocene vertebrates of Indiana.

Proceedings of the Indiana Academy of Science **93**:483–504.
Richards, R. L. (1990). Quaternary distribution of the timber rattlesnake (*Crotalus horridus*) in southern Indiana. *Proceedings of the Indiana Academy of Science* **99**:113–22.
Richards, R. L., Whitehead, D. R., and Cochran, D. R. (1987). The Dollens mastodon (*Mammut americanum*) locality, Madison County, East Central Indiana. *Proceedings of the Indiana Academy of Science* **97**:571–81.
Richmond, N. D. (1964). Fossil amphibians and reptiles of Frankstown Cave, Pennsylvania. *Annals of Carnegie Museum* **36**:225–28.
Rickart, E. A. (1977). Pleistocene lizards from Burnet and and Dark Canyon caves, Guadalupe Mountains, New Mexico. *Southwestern Naturalist* **21**:519–22.
Rogers, K. L. (1982). Herpetofaunas of the Courland Canal and Hall Ash local faunas (Pleistocene:Early Kansan) of Jewell Co., Kansas. *Journal of Herpetology* **16**:174–77.
Rogers, K. L. (1984). Herpetofauna of the Big Springs and Hornet's Nest quarries (Northwestern Nebraska, Pleistocene:Late Blancan). *Transactions of the Nebraska Academy of Sciences* **12**:81–94.
Rogers, K. L., Repenning, C. A., Forester, R. M., Larson, E. E., Hall, S. A., Smith, G. R., Anderson, E., and Brown, T. J. (1985). Middle Pleistocene (late Irvingtonian) climatic changes in south-central Colorado. *National Geographic Research* **1**:535–63.
Roth, J. A., and Laerm, J. (1980). A late Pleistocene vertebrate assemblage from Edisto Island, South Carolina. *Brimleyana* **3,** 1–29.
Saunders, J. J. (1977). Late Pleistocene vertebrates of the western Ozark highlands. *Illinois State Museum Reports of Investigations* **33**:1–118.
Savage, D. E. (1951). Late Cenozoic vertebrates of the San Francisco Bay region. *University of California Publication of the Department of Geological Sciences Bulletin* **28**:215–314.
Sellards, E. H. (1916). Fossil vertebrates from Florida; a new Miocene fauna; new Pliocene species; the Pleistocene fauna. *Florida Geological Survey Annual Report* **7**:121–60.
Slaughter, B. H. (1967). Animal ranges as a clue to late Pleistocene extinction, In P. S. Martin and H. E. Wright, eds. *Pleistocene Extinctions: The Search for a Cause*. New Haven: Yale University Press, pp. 155–67.
Steadman, D. W., and Martin, P. S. (1984). Extinction of birds in the late Pleistocene of North America. In P. S. Martin and R. G. Klein, eds. *Quaternary Extinctions: A Prehistoric Revolution*. Tucson: University of Arizona Press, pp. 466–82.
Stuart, A. J. (1979). Pleistocene occurrences of the European pond tortoise (*Emys orbicularis* L.) in Britain. *Boreas* **8**:359–71.
Stuart, A. J. (1982). *Pleistocene Vertebrates in the British Isles*. London: Longman.
Stuart, A. J. (1991). Mammalian extinctions in the late Pleistocene of northern Eurasia and North America. *Biological Reviews* **66**:453–562.
Sutcliffe, A. J. (1985). *On the Track of Ice Age Mammals*. London: British Museum (Natural History).
Tanner, W. W. (1978). Zoogeography of reptiles in the intermountain region. *Great Basin Naturalist Memoirs* **2**:43–53.
Taylor, E. H. (1943). An extinct turtle of the genus *Emys* from the Pleistocene of Kansas. *University of Kansas Science Bulletin* **24**:249–54.
Tihen, J. A. (1942). A colony of fossil neotenic *Ambystoma tigrinum*. *University of*

Kansas Science Bulletin **28**:189–98.

Tihen, J. A. (1952). *Rana grylio* from the Pleistocene of Florida. *Herpetologica* **8**:107.

Tihen, J. A. (1954). A Kansas Pleistocene herpetofauna. *Copeia,* no. 3 pp. 217–21.

Tihen, J. A. (1958). Comments on the osteology and phylogeny of ambystomatid salamanders. *Bulletin of the Florida State Museum, Biological Sciences* **3**:1–50.

Tihen, J. A. (1960). Notes on late Cenozoic hylid and leptodactylid frogs from Kansas, Oklahoma and Texas. *Southwestern Naturalist* **5**:66–70.

Tihen, J. A. (1962). A review of New World fossil bufonids. *American Midland Naturalist* **68**:1–50.

Tihen, J. A. (1964). Tertiary changes in the herpetofaunas of temperate North America. *Senckenbergiana Lethaea* **45**:265–79.

Van Dam, G. H. (1978). Amphibians and reptiles (in the Baker Bluff Deposit, Tennessee, and the Late Pleistocene Faunal Gradient). *Bulletin of the Carnegie Museum of Natural History* **ll**:19–25.

Van Denburgh, J. (1895). A review of the herpetlogy of lower California. Part I: reptiles. *Proceedings of the California Academy of Sciences* **5**:77–162 plus 14 plates.

Van Devender, T. R., and King, J. E. (1975). Fossil Blanding's turtles, *Emydoidea blandingii* (Holbrook), and the late Pleistocene vegetation of Western Missouri. *Herpetologica* **31**:208–12.

Van Devender, T. R., and Mead, J. I. (1978). Early Holocene and late Pleistocene amphibians and reptiles in Sonora Desert packrat middens. *Copeia,* no. 3 pp. 464–75.

Van Devender, T. R., Moodie, K. B., and Harris, A. H. (1976). The desert tortoise (*Gopherus agassizii*) in the Pleistocene of the northern Chihuahuan desert. *Herpetologica* **32**:298–304.

Van Devender, T. R., Phillips, A. M., III, and Mead, J. I. (1977). Late Pleistocene reptiles amd small mammals from the lower Grand Canyon of Arizona. *Southwestern Naturalist* **22**:49–66.

Van Devender, T. R., and Tessman, N. T. (1975). Late Pleistocene snapping turtles (*Chelydra serpentina*) from southern Nevada. *Copeia,* no. 2 pp. 249–53.

Van Devender, T. R., and Worthington, R. D. (1977). The herpetofauna of Howell's Ridge Cave and the paleoecology of the northwestern Chihuahuan desert. In R. H. Wauer and D. H. Riskind, eds. *Transactions of the Symposium on the Biological Resources of the Chihuahuan Desert Region, United States and Mexico.* U.S. Department of the Interior National Park Service Transactions Series No. 3, pp. 85–106.

Voorhies, M. R. (1987). Fossil armadillos in Nebraska: the northernmost record. *Southwestern Naturalist* **32**:237–43.

Weaver, W. S., Jr., and Robertson, J. S. (1967). A re-evaluation of fossil turtles of the *Chrysemys scripta* group. *Tulane Studies in Geology* **5**:53–66.

Webb, S. D., ed. (1974). *Pleistocene Mammals of Florida.* Gainesville: University of Florida Press.

Weigel, R. D. (1962). Fossil vertebrates of Vero, Florida. *Florida Geological Survey Special Publication* **10**:1–59.

Wilson, V. V. (1975). The systematics and paleoecology of two Pleistocene herpetofaunas from the southeastern United States. Ph.D. dissertation, Michigan State University, East Lansing.

Woodburne, M. O., ed. (1987). *Cenozoic Mammals in North America, Geochronology and Biostratigraphy.* Berkeley: University of California Press.

INDEX

This index relates to the chapters in the book other than Chapters 4 and 6 which are covered by the site and taxonomic indices.

SITE INDEX

North American Pleistocene herpetological sites (exclusive of Mexico) are listed alphabetically in the order of the regions designated in this book. Abbreviations: IRV= Irvingtonian Land Mammal Age; RLB= Rancholabrean Land Mammal Age. (United States Postal Service Standard Abbreviations are used to designate states.)

Great Lakes Region

Appalachian Region

Southeastern Region

Ozark Region

Northern Plains Region

Southern Plains Region

Rocky Mountain/Great Basin Region

Pacific Coast Region

TAXONOMIC INDEX

North American Pleistocene genera (exclusive of Mexico) mentioned in the systematic and faunal list sections of the book are listed here in alphabetical order.